Front Cover Photographs: Ralph Sarich with the 1988 X-series OCP engine. Left Inset: The Orbital engine. Right Inset: The OCP fuel-injection system as applied to a marine engine.
Rear Cover Photograph: Ralph Sarich demonstrates his Orbital Engine to Renault executives in late 1973.

SARICH

THE MAN AND HIS ENGINES

RECENT BOOKS BY PEDR DAVIS

Wheels Across Australia — Australian Motoring
From the 1890s to the 1980s

The Macquarie Dictionary of Motoring
(with Tony Davis)

Charles Kingsford Smith — The World's Greatest Aviator
(new edition with foreword by Nancy-Bird Walton)

Which Car For You? — The Australian Buyer's Guide
To Used Cars (Third Edition)

Project VN — An Australian Car for the 1990s
(with Tony Davis)

SARICH

THE MAN AND HIS ENGINES

BY PEDR DAVIS

MARQUE
PUBLISHING COMPANY

Distributed by Universal Press
and
Gregory's Scientific Publications

SARICH – THE MAN AND HIS ENGINES

This book was written by Pedr Davis and researched by Pedr Davis and Tony Davis. It was first published in 1989 by MARQUE PUBLISHING COMPANY PTY LTD, PO Box 203 Hurstville NSW 2220 Australia.

National Library of Australia:

Davis, Pedr
Sarich – the man and his engines.

Includes index.
ISBN 0 947079 08 4.

1. Sarich, Ralph, 1938- . 2. Inventors – Australia – Biography.
3. Engines. I. Title

Proudly produced wholly within Australia
Design and production by Tony Davis
Cover layout by Irene Meier
Copy editing by Anne Sahlin
Mono photo-screenings by Terry Clark Typesetters Pty. Ltd.
Typesetting by Terry Clark Typesetters Pty. Ltd.
Assembly by Type Forty, Glebe
Printed by Macarthur Press, Sydney

Distributed by Universal Press and
Gregory's Scientific Publications.

ACKNOWLEDGMENTS

It is almost impossible to write a non-fiction book without borrowing expertise from a number of people and I've been fortunate in having an unusually talented team to lean on.

My special thanks to Marque stalwarts Tony Davis, Anne Sahlin and Irene Meier for contributing their customary highly professional services. Ian Porter, who knows far more about the world of high finance than I, made a valuable contribution in his specialised area. Barry Lake, the enthusiastic editor of Modern Motor magazine, not only loaned some historic photographs to round out the Davis collection, his journal was the first to publish my original evaluation of the OCP two-cycle technology. The Sunday Times in Perth was also generous in making photographs available.

Although they had no way of knowing what kind of book would emerge, I had nothing but help and cooperation from the dedicated group at the Orbital Engine Company. Firstly, Ralph Sarich was very generous with his time and helpful in digging into the past to answer my queries. I am also grateful to OEC personnel Ken Johnsen, Kim Schlunke, Tony Fitzgerald and — before he left the company — John Cook. Mrs Patricia Sarich kindly granted me one of the very few interviews she has ever given to the media.

Valuable assistance was also provided by Kathleen McMahon, Denis Creer and Peter Walsh BA LLM.

Pedr Davis
Sydney
February 1989

'He is like a computer — you can almost see him thinking. When you have a problem you are usually satisfied if you can get one solution. Sarich comes up with 20 or 30.'

Dr Robert Ward, BHP, 1973

'I get a lot of self-satisfaction trying out my ideas. It's a matter of wanting to achieve something in life. I want to achieve something extraordinary.'

Ralph Sarich, 1987.

Contents

Perspective

For most of his life, Ralph Sarich has been obsessed with designing an engine which would change the world. He set himself the task of producing a light, simple, fuel-efficient unit which would comply with emission standards.

He started work in a small workshop in an outer suburb of Perth in Western Australia and — in his first attempt — built a rotary-type power plant, which he named the Orbital engine. This created press hysteria but never went into production. The second engine, the Orbital Combustion Process (OCP) design, met with a very subdued — even a disparaging — press reaction but is now the darling of the motor industry.

Sarich himself has become one of Australia's wealthiest, best-known and most controversial figures.

The controversy was aroused by a prolonged period of media hype, false starts, severe setbacks and furious work. But, unfairly, he remained controversial even after producing an engine which is considerably smaller, lighter, more economical, more powerful and far less polluting than any production unit of similar capacity ever built.

The design has taken Detroit, Tokyo and Europe by storm and should earn the Sarich organisation — and Australia — untold wealth.

Despite this success, Sarich remains the target of public and private sniping. He has been called a conman, a technical ignoramus, a sensational publicity-seeker and the undeserving beneficiary of business grants and taxpayers' money.

His critics charge that after two decades of development he still does not have the engine in mass production. It is even claimed that Rudolf

Diesel invented the two-stream fuel-injection system which is the lynchpin of Sarich's OCP design.

What are the facts?

This book is designed to explore the fascinating story and sift fact from fantasy. No doubt, some of the early criticism was justified. But when the former fitter-and-turner emerged from a backyard operation to become a world authority on engine technology, he confounded his critics. Many have not forgotten or forgiven him. As the winner of the 1988 Churchill Medal, awarded by the British-based Society of Engineers, Sarich takes his place beside Sir Frank Whittle of jet-engine fame and Sir Christopher Cockerell who invented the Hovercraft.

For the first time, an outsider examines the strengths and weaknesses of the Orbital engine which created so much media hysteria and explains how work on this original design led to Sarich creating one of the world's foremost engine research organisations.

This biography was researched and written over a period of two years, aided by some exclusive interviews with Ralph Sarich and his wife, Patricia, as well as Orbital Engine Company staff and some media critics. It tells of Sarich's remarkably original approach to engineering problems, his fierce determination to succeed and his extraordinary negotiating skills.

Sarich — The Man and His Engines — also reveals why his first design, the much-publicised Orbital unit, never went into production and why the technology behind the current design will radically change the shape of cars for the 1990s.

The technology developed by Sarich and his team is so advanced that several major car companies have signed licensing and other agreements which mean that, for the first time in their histories, they will use engines built by or licensed from an outside organisation.

The book reveals that, despite claims to the contrary, Sarich has not personally received any government money. His companies have been granted a total of $6.1 million in State and Federal government funds — a figure less than some car companies receive from the Australian government in a single year for their research work. Sarich's Orbital Engine Company has also received approximately $17 million from The Broken Hill Proprietary Company Ltd (BHP), the Australian steel-making conglomerate, which is an equal partner in the operation. This was not a grant but payment for the acquisition of an equity in the business.

The $24 million spent developing both engines and the fuel-injection system is a small sum by industry standards. Most completely new versions of conventional mass-produced engines cost considerably more to design and develop. Even a new cylinder head can cost $10 million while some experimental engines have run up bills amounting to several hundred million.

The time factor is also small by industry standards.

Sarich spent ten years developing the rotary-type engine which has now been put aside in favour of a more conventional reciprocating unit designed around his unique combustion process. Work on this second engine started in 1981 and was substantially finished eight years later. In contrast, the original diesel engine took 32 years to reach mass production and the Wankel rotary nearly 40 years.

It is true that Rudolf Diesel and others experimented with two-stream fuel-injection, but they were not able to make it work. Sarich found the answer and has worldwide patents on a fundamentally different system which bears little resemblance to the crude technology employed by Diesel.

I hope that this book leads to a wider understanding of the man and his motives. I especially want the world to see the scope of the technology achieved by Ralph Sarich and his hardworking team.

Even at this late point, fate could step in and bring progress to a halt - but this is improbable. The Orbital Engine Company is travelling in the fast lane. Negotiations, licences and decisions are now moving rapidly and present a constantly changing target for anyone reporting the facts.

The last chapter of this book was finalised in January 1989. Between then and the day it reaches the bookshelves, many things will have changed. But the basic technology, the long, long years of hard work and the effect the man from Western Australia has had on the world's automobile industry is a matter of history.

Pedr Davis
February 1989

SPECIAL NOTE: Because the original orbiting and the later reciprocating engines employ similar combustion technology, Sarich calls both Orbital units. To avoid confusion, the name Orbital engine in this book refers to the orbiting piston type, while OCP engine refers to the reciprocating type.

1

Turning the auto industry on its ear

Inventors have been looking for the golden egg ever since the automobile ceased to be a rich man's plaything.

The arithmetic is simple.

Last year, the world's car makers built 45 million motor vehicles. By producing even a humble component earning a ten cent profit margin, an inventor could reap vast riches if it was fitted to all cars.

What then of a man who has patented engine technology which stands a better than even chance of being adopted not just by the auto industry but by the vast market for outboard and inboard marine engines, motorcycles, light aircraft and industrial power plants?

Ralph Sarich from Perth, the capital city of Western Australia, has signed contracts with several major companies which could — and most probably will — lead to his technology being used in a sizeable proportion of the 100 million internal combustion engines built each year.

By the end of 1988, Sarich had signed full licensing agreements with Ford, Outboard Marine, Mercury and Walbro Corporation. As we go to press, he is in Detroit negotiating with General Motors (GM) while several Japanese firms wait to pay their respects and get down to hard bargaining.

And I mean hard.

Sarich knows the value of his invention. He will not talk money to outsiders, but when I suggested an average royalty figure of US$40 per engine, he smiled and nodded as though agreeing it was a good guess. By mass-production standards, $40 per engine is a huge sum but independent studies show that the Orbital Combustion Process (OCP) engine can be built much more cheaply than any existing automotive engine and there are flow-on cost savings because of the reduced weight. It will also cut down the owner's fuel bill, provide other benefits and reduce the exhaust emission – so companies are prepared to pay the fee.

Sarich was already financially well placed when he won an ABC TV award for inventing a new kind of engine back in 1972. At the time, newspaper headlines claimed the engine would revolutionise the motor industry. The prophesy proved false but this engine, which Sarich called an Orbital design, led to the development of new combustion technology which indirectly made him one of Australia's four richest men.

If his new engine makes the inroads into the automotive industry that he and many industry people expect, he is destined to become one of the wealthiest Australians ever.

Ralph Sarich was 32 years old in 1970 when he decided to quit his job selling heavy earth-moving equipment and start a small auto research business. The sheer audacity of the scheme did not occur to him. Each year, GM alone is offered 30,000 new inventions, and after preliminary sortings, they fully investigate only two or three of them. The rest mainly languish in that great stockpile of fallen monuments and broken dreams. The chances of commercial success – even for an inventor with brilliant new technology – is remarkably low. For an untrained, unknown engineer living in Western Australia, the chance of success must have bordered on zero.

To compound the improbable, Sarich chose the worst possible kind of invention for his initial attempt to crack what little market exists for outside inventions in the automotive field. He wanted to sell the industry a revolutionary rotary engine. The history of such inventions has been one long financial catastrophe and nothing Sarich had in mind seemed likely to change the situation.

Inventors hoping to reap vast rewards from the auto industry often pin their hopes on a variation on the rotary concept. The theory is promising. In a rotary engine, the piston spins continuously in one direction, instead of reciprocating up and down a cylinder as in a conventional petrol or diesel engine. To make the rotary idea work, it is necessary to devise a simple and efficient means of varying the size of the combustion chamber to enable the fuel mixture to be compressed, ignited and exhausted after the piston has drawn it in. If this can be done effectively, a rotary engine should have some overwhelming advantages over the conventional unit.

In theory at least, a well-designed rotary should be smoother, smaller, lighter and more powerful than a reciprocating engine of comparable output. It needs fewer parts and, because of its lightness, would reduce the need for power steering and a heavy-duty front suspension.

An estimated 1500 different rotary engines have been patented over the years but only the Mazda-built Wankel has been financially successful. Even in this case, the high cost of development almost sent the company broke before the engine became competitive. Today, 20 years after its first production rotary car was announced, Mazda sells far more cars with conventional engines than Wankel-powered units.

In 1970, when Ralph Sarich decided to get into the act, no one — not even Mazda — had built a commercially successful rotary. In fact, car companies were soon to become extremely wary of spending development money in this area because of what came to be termed the 'Wankel affair'.

Twelve years earlier, after 30 years work, Dr Felix Wankel of Germany had announced a novel form of rotary engine which was launched with considerable hype. Widely acclaimed as a sensation, the engine brought rave notices from the cognoscenti, many of whom said it would make the conventional piston engine obsolete.

Already closely associated with the German car and motorcycle maker NSU (now a division of Volkswagen), Dr Wankel went into partnership with that firm with a view to making rotary engines suitable for a wide range of transport.

The novel engine was based on a Wankel-designed rotary air compressor which had been used as a supercharger for an NSU record-breaking motorcycle in 1952.

Felix Wankel, who died in 1988 at the age of 86 years, had much in common with Sarich. The son of a forestry worker, he went to work before finishing school and became a professional engineer at the age of 24 years without the benefit of a university education. His main passion was his work on rotary valve motorcycle engines until about 1930 when he designed an unusual rotary unit. He applied for patents for it in 1933, when aged 31 years. It was not until 1950 that NSU agreed to finance the development work through a new company, Wankel-NSU. The original Wankel engine was such that the entire unit rotated, but after the initial tests, NSU engineers (who took on the main development task) started work on a new design. This unit, designated a circulating piston engine, has the outer housing stationary with an inner rotor (or piston) making a planetary circulating motion. During each rotation, the rotor creates a number of sealed chambers, each of which steadily varies in size. Fuel mixture is drawn into these chambers where it is compressed, ignited, allowed to expand and then leave via the exhaust system. The intake and exhaust ports are fixed in the housing, as is the spark plug.

The prototype of the Wankel original design weighed only 11 kg and developed 22 kilowatts at 17,000 rpm. It created worldwide interest and was featured throughout the world's media as the 'engine of the future'.

NSU immediately got to work designing a car suitable for the Wankel rotary – the Spider Sports, released in 1964. The Spider was followed in 1967 by the NSU Ro80 sedan which was intended for large-scale production. The advanced-looking, front-wheel drive car was powered by a twin-cylinder rotary engine. This developed 85 kW yet weighed half as much as other car engines of similar output. Marine and other variants of the Wankel engine were also designed.

Impressed by its potential, Curtiss-Wright Corporation of the US took out a licence and agreed to devote a large proportion of its research budget to developing the Wankel concept for the US market. It had in mind light aircraft, trucks, buses, portable generators and other power plant applications, as well as the huge potential market for cars. Licences were taken out by such prestigious companies as Rolls-Royce, Mercedes-Benz, Citroen, Mazda and GM. Even the Russian motor industry built prototypes for use in future Volga models.

Several major firms spent large sums designing and developing Wankel engines. Like NSU, they ran into severe durability problems, mostly centred around the rapid wear of the seals which keep the burning gases within the combustion chambers. Engineers also ran into major fuel consumption and exhaust emission problems. Despite the difficulties, Citroen built a large plant and started manufacturing Wankel-powered cars. The plant closed less than a year later, due to unexplained technical problems. In due course, Curtiss-Wright, Rolls-

Royce and Mercedes abandoned their costly research on Wankel engines. NSU, Mazda and GM persevered and in 1967 Mazda released the rotary-powered Cosmo Sports coupe. Even here, the technical difficulties became so acute that the firm spent vast sums overcoming them.

NSU continued to make Ro80 cars and, at one stage, was giving free replacement engines to owners who ran into trouble. The cost of the development program was so large it drove the firm into serious financial difficulties. In 1969 NSU amalgamated with Audi and was later taken over by Volkswagen.

Mazda felt it was worthwhile battling against the never-ending technical problems and later produced a commercially viable Wankel engine. By 1988, the Japanese company had built nearly two million rotary units, but was still selling eight conventional engines for every rotary.

In financial terms, no one was hit harder by the Wankel concept than GM. The giant US corporation had shown immense confidence in the Wankel and publicly stated its commitment to mass-produce a line of rotary-powered cars. In contrast, Ford and Chrysler tended to downplay the Wankel's likely impact.

GM embarked on a new program to develop the Wankel in 1970 with a view to launching a rotary-powered Chevrolet Vega in late 1975. The Vega was to be followed by several other Wankel-powered models. At the height of the development program, GM had 200 specialists working on rotary units in the corporation's Technical Centre in Warren, Michigan. They reported to GM's president Edward N. Cole who assumed full responsibility for switching to the new power plant. Another group worked on Wankel assignments at Chevrolet Engineering and yet another at GM's Hydramatic Division in Ypsilanti, Michigan.

GM paid about US$42 million for licensing agreements and, by the time the project had been aborted in 1977, had spent an estimated $700 million on engineering and production machinery. Company insiders said that the program was abandoned because the company could not meet the durability, fuel consumption and exhaust emission targets needed to make the unit competitive. The compact size, smoothness and low weight made it an attractive proposition but the drawbacks — and high cost of production machinery — proved insurmountable.

Even in 1970, the same year that Ralph Sarich planned to enter the engine business, some engineers had begun questioning the Wankel's real potential. But this did not deter Sarich from conceiving yet another unconventional unit, one with an unusual motion in which the rotor/piston revolves in an orbital fashion and not about its own axis. To distinguish it from a rotary unit (in which the rotor/piston spins about its axis), Sarich named his design an Orbital engine. The Australian media took up the cause in full cry, at times making the most outrageous claims about the engine's advantages and how it had taken the world by storm. No other Australian invention before or since has received so much media attention. Press stories about huge sums being offered for the sole manufacturing rights were not fabricated, however. Sarich says at least one multimillion dollar offer was made.

Much of the publicity was generated in 1972 after Sarich apppeared on the ABC television program 'The Inventors' where the engine made its public debut driving a small generator. The producers of the program had sought the advice and opinions of several outside organisations, including the Royal Automobile Club of WA, and had seen the engine in operation. However, Sarich was not asked to provide such technical details as the torque output, operating speed, specific fuel consumption or other data which engineers would expect as a matter of routine.

He was acclaimed by the ABC as the 'Inventor of the Year'.

'About a year after I appeared on The Inventors program', said Sarich in 1988, 'some guy representing a group of people approached me. Initially he offered me $12 million for the rights. I declined and he kept going until he reached $18 million. He put the offer in writing and wanted to tie me down to a long-term contract developing the engine.

'I was never really tempted because I wanted to do my own thing. If you talk to Pat [Ralph's wife] she'll tell you that we spent only a few minutes talking about it and then agreed to break off further negotiations.'

There must have been times later when he regretted this decision. What Sarich did not realise at the time is that the auto business moves very, very slowly. Risks are rare, especially when it comes to unconventional engineering. No matter how good it may be, the prospects of a rotary engine making the big time within a few years of conception were — and still are — extremely slim. No major firm would dare stake its reputation for building reliable and durable motor cars by introducing an engine which had not been proven to the last nut and bolt. If the major car firms have a collective fault, it is that they err on the side of overcaution.

'Had I been aware of what I know today, I would have concluded that it was almost impossible to convince the ultraconservative auto industry', Ralph Sarich told this writer some years later.

'But I was confident that if you dedicate yourself and commit yourself to a task, you must succeed.

'The Orbital was the first engine I built but I had looked at numerous geometries before settling on it in 1970. The main appeal was the compact size and light weight but I also wanted a piston motion which would be conducive to developing my stratified charge concept. The Orbital geometry did this by physically dividing the combustion chamber into numerous sections as the piston approached the top dead centre and went past it.'

It took two years, $60,000 and the help of a few committed friends to get the engine running. Several variations were made and eventually one was installed in a Ford Cortina for testing. The engine was promising, but no more so than the Wankel and dozens of others had been.

Inspired by the ABC television program, BHP — Australia's largest company — agreed to back the Orbital concept in late 1972 to a maximum figure of $50 million. Twelve years later, Sarich launched a public company — Sarich Technologies Trust, now Sarich Technologies Ltd (STL), whose shares have a current (1989) market value around $350 million. STL and BHP are joint owners of Orbital Engine Company Pty Ltd, the operating company devoted to engine combustion research and development.

Despite the fact that the Orbital engine had received virtually no on-road development, Sarich persuaded GM to spend hundreds of thousands of dollars investigating it several years after the corporation had lost heavily on its Wankel project. GM took the decision following a discussion between Sarich and Howard Kehrl, a senior GM vice-president. In 1980, Sarich provided dynamometer readings and came close to delivering to GM a car complete with an Orbital engine. His dyno figures showed that his engine had a big weight saving over the Buick V6, used less fuel and came within 5 per cent of its emission figures, being better in some areas, not quite as good in others.

After some discussions, GM sent a V6 — then a state-of-the-art engine — so Sarich engineers could put it through its paces on their own dynamometers. GM wanted to be sure the test facilities in Perth were not optimistic and spoke the same language as those in Detroit. They were agreeably surprised to find that the Sarich print-outs not only duplicated the GM readings but provided additional information. It was the impressive performance of this rotary engine which later made GM interested in the second Sarich engine. This is a reciprocating three-cylinder, two-cycle unit which now promises to revolutionise the auto industry to a far greater extent than would have been possible with the first.

The second engine came about when Sarich was trying to improve the Orbital engine's combustion process and reduce the exhaust emission levels via a stratified charge. He had earlier developed a novel form of two-stream fuel-injection which was tried unsuccessfully on his first Orbital engine, and later he refined the idea and overcame the earlier problems.

The new combustion system was first tested with the Orbital engine and worked well. It was also tried in a conventional four-cycle piston engine installed in an Australian Holden Camira in 1982. Here it demonstrated a fuel economy advantage around 25 per cent and, more importantly, comfortably met Australian legislated emission levels.

The novel injection is called a two-stream system because it uses compressed air to carry a measured quantity of fuel directly into each cylinder. Combined with a variable exhaust valve, it probably represents the most advanced combustion technology available today. It is the key to the amazing interest which the world's auto industry has shown in Sarich's operation since 1984.

Let there be no doubt that the interest has been extraordinary. Before Ford had signed its licensing agreement, an engineer from a famous European manufacturer flew to Perth, inspected the operation and returned home. A few weeks later, a surprised Sarich showed me a letter, written on private notepaper, in which the same engineer asked if he could buy shares. Another engineer, representing a British firm, came to Perth and wanted to know if Sarich could develop a 12-cylinder version for a luxury car planned for the 1990s.

The success of the two-stream injection system proved the turning point for Sarich and the company. The system worked well on the Orbital engine and OEC engineers determined that the gas motion of a two-cycle engine could be exploited by a similar process provided that some basic changes were made to the engine's design. Sarich authorised experimental work to be carried out and it met with instant success. The results were presented to certain automobile companies and their response was sufficiently encouraging for Sarich to recommend to the OEC board that work be suspended on the Orbital design in favour of refining the injection process for conventional reciprocating engines. The board agreed that the concept would be far easier to sell to car makers than the rotary engine because it could be manufactured on existing equipment and would not be seen by potential customers as extreme or unproven.

The system is suitable for conventional four-cycle reciprocating engines and for two-cycle designs which are currently widely used for motorcycles and boats but seldom for cars. Sarich chose to develop the latter because a variety of technical advantages follow and these are described later. It is enough to say here that the concept has several major advantages when applied to a two-cycle engine. Comparing the engine with a conventional four-cylinder unit of comparable power built by Ford or GM, the Orbital Engine Company (OEC) lists the advantages as follows:

* The engine is one-third the weight and occupies one-quarter the packaging space.
* It uses about 30 per cent less fuel. The standard vehicle returns 40 miles per gallon (7 litres per 100 kilometres), the Sarich version 51 mpg (5.5 l/100 km) when subject to the same US EPA urban driving cycle test.
* According to OEC engineers, the Sarich unit develops 78 kW (105 bhp gross) from a capacity of 1.2 litres and has a torque curve better than a very good four-cycle production engine. Further development will include a forced induction version developing 97 kW (130 bhp), using a novel, low-cost supercharger.
* The untreated emission levels are five times lower than for any two-cycle engine of its power ever built. The hydrocarbon emissions are sensationally below the current US limit, nitrous oxides are 50 per cent lower and carbon monoxide is cut by 92 per cent. These levels have been achieved with a simple oxidising catalyst, not the expensive triple bed design used in some cars.

* Because it is a three-cylinder, two-cycle unit without a valve train and associated parts, the OCP engine requires about 250 fewer parts and is about 30 per cent cheaper to manufacture than a four-cylinder, four-cycle unit of equivalent power.
* It can be produced with current production equipment and should not create durability problems.

As an experienced automotive engineer and a fairly hard-nosed technical journalist, this writer believes the above claims. I've taken a close look at the operation, watched a number of engines being tested on dynamometers and seen the test results. I've also driven Ford and GM cars fitted with Sarich engines.

But the real proof comes from Detroit where the unit has been exhaustively tested by Ford and GM. OEC targeted GM as it is the world's largest vehicle maker. When Ralph Sarich first tried to interest GM, he met the expected resistance. Novel engines are a dime a dozen and spectacular claims come as freely as press releases. GM had large reservations about Sarich's ability to meet future emission legislation. So Sarich said: 'These are the advantages. We will install our engines in your cars, and if they do not achieve our stated claims, there will be no charge. If they meet the claims, you pay an agreed fee.'

'Even this offer was not seized upon eagerly by GM', Sarich recalled in 1988. 'This type of proposal was novel in their history and they seemed unsure how best to tackle it. But finally they agreed.'

GM then sent two cars to Perth where Sarich fitted the engines. One car stayed in Perth for further testing, the other was taken to Detroit by an OEC senior engineer who could answer any questions which arose. The car was tested for eight months, then GM paid Sarich in full, acknowledging that all targets had been met.

By the end of 1987, BHP's capital injections were completed and Orbital was beginning to see income from its licensing activities. The shortfall in income to cover the operational cost was funded by borrowing. By that time, BHP had provided about $17 million for their equity in the business and the Federal and Western Australian governments have put in $6.1 million between them.

Over the years, the financial market has reacted strongly to every whisper and every hint that Sarich was reaching his goals. And no wonder. The aim is to have Sarich technology in every engine built anywhere. The potential market cannot be overstated. The technology could be applied to the 45 million motor vehicles, 5 million motorcycles, 2.2 million marine installations and large numbers of industrial, aeronautic, lawnmower and power generators built each year.

Motor vehicles are already subject to stringent emission controls — and similar legislation is on the way for marine engines and other forms of motive power. Expensive and complex equipment is required to meet the emission levels but Sarich technology provides the most simple and least expensive answer yet. This is why the world is beating a path to his door.

His technology has demonstrated exceptionally low emission levels for a wide variety of piston engines in addition to reduced weight, package size and fuel consumption. It is suitable for any petrol engine from 50 ccs upwards. Ideally, the new technology should be used in a two-stroke engine in multiples of three cylinders because an optimum exhaust configuration can be used.

Many automotive engineers agree that piston engines should operate on the two-cycle principle. Such units have been used in DKW, Saab, Suzuki and Wartburg passenger cars as well as in countless diesel trucks. Two-stroke engines are smaller, lighter and cheaper to make than four-cycle units of comparable power. They also have reduced internal friction, partly because the crankshaft does not slosh around in a sump of oil, but also because rolling element bearings are used and power-wasting oil scraper rings and valve train mechanisms are not needed.

However, very few cars powered by a two-cycle engine have been commercially successful. Several major car companies have sunk large sums into development programs to produce hassle-free two-cycle engines only to find they cannot get good torque output at low engine speeds, acceptable fuel economy and low exhaust emissions.

Thanks to a radically different combustion process described later, the Sarich technology overcomes the problems traditionally associated with two-cycle engines and this is why Ford, GM and two Japanese firms have new small cars on the drawing boards, powered by Sarich-inspired two-cycle engines.

At the time of writing Chrysler Corporation is heavily involved in the development of its own promising supercharged, two-cycle engine for passenger cars but reports from Detroit indicate that its power-to-weight ratio is well below that of the OEC design. Interestingly, Chrysler resumed discussions with the Sarich team in 1989.

The booming marine market has also turned to Sarich. Between them, Outboard Marine Corporation and Mercury — who have signed licensing agreements with Sarich - build over a million engines per year. Sarich is also negotiating with several Japanese marine engine manufacturers attracted by the low emission levels, better starting, improved idle stability, absence of smoke and superior fuel economy.

There are further advantages for car makers. The low packaging size means that cars could be restyled for better aerodynamics and more usable space. Ford has already shown how this could be done in the experimental Saguaro displayed at the 1988 Geneva Motor Show.

As this car's shape shows, a very small engine allows a completely new approach to the bonnet angle and front suspension design, with the spare wheel and other components fitting under the bonnet. The greatly reduced engine weight means there's less need for power steering and other flow-on advantages affect the entire vehicle. As a rule of thumb, a 1 kg reduction in engine weight leads to a 2 kg reduction in overall vehicle weight.

Unlike most other novel engines, including the Wankel, the Sarich design can be mass-produced using existing tooling. The simplicity means a big reduction in manufacturing cost, even after paying a substantial royalty. Kim Schlunke, now the Sarich company's chief engineer, estimates the direct savings at $300 per engine compared with a four-cylinder car engine of comparable power.

In contrast to some radical propositions, including the original Orbital design, there should be no durability problems. Other than the direct injection and unusual combustion system, the unit follows established two-cycle principles. Marine engine companies say they have field data which shows that two-cycle engines are more durable under heavy loads than comparable four-cycle designs.

The old two-stroke bogies of smoke and high oil consumption have been eliminated. The engine runs on regular unleaded fuel with a separate oil supply and uses about the same amount of oil as a standard four-cycle engine given regular oil changes. It has a plug-in replacement oil cartridge.

No one knows who will be first to produce a production car powered by a Sarich-inspired engine, but the betting is on Ford. The show-stopping Saguaro displayed in the 1988 Geneva Motor Show is only one advanced prototype built for a new generation of compact engines. Some insiders believe that a Sarich engine will initially be offered as an option in a baby car which will replace the current Ford Fiesta in the early 1990s. GM is not far behind in its development program and at least two Japanese firms — and one Korean — are dark horses to launch the world's first production car with a Sarich-patented engine.

Provided that the public accepts the concept — and there's no reason not to, other than a psychological objection to a two-stroke based on previous poor designs — the engine will be used increasingly in future Ford models.

Other firms will be forced to follow.

Although the concept has most to offer as a three-cylinder unit powering a relatively small car, it can be used as a six-cylinder or 12-cylinder engine in family and luxury models. Several major companies have now expressed keen interest in this area and Sarich has a supercharged V6 version underway in his Balcatta workshops.

Says Kim Schlunke, chief engineer at Orbital Engine Company: 'I believe our combustion process will be universal technology in a short period of time'.

If he is right, then Sarich, the unqualified engineer from Western Australia, will have pulled off the biggest feat in the history of the automobile. No one else has ever drawn royalties on a major item of technology used by all major car companies.

The closest has been another Australian, Sydney's Arthur Bishop, who patented the variable ratio power steering system now fitted to cars made in Japan, the USA and Europe.

2

Orbital days of hope and hype

Perth is the gentle, picturesque home to a million West Australians. Founded in 1829, it is one of the prettiest locations imaginable, being built around the Swan River, overlooking the Indian Ocean and nesting in the foothills of the Darling Range. As the capital city of Australia's largest State, it is the financial headquarters of one of the greatest mineral treasure chests to be found anywhere in the world. Several decades of booming mining ventures have turned Perth into a wealthy and sophisticated city but these are facts known to relatively few Australians.

Over 4000 kilometres of desert and bush separate Perth and Australia's heavily populated eastern seaboard. Despite the modern communication network, a great deal can happen in Western Australia and not reach the rest of the country. So it was with Sarich fever, at least initially. High temperatures were running in the West from 1971 onwards but the eastern media showed no interest until mid-1972 when the Australian Broadcasting Commission, as it then was, flashed Sarich's face and Orbital engine across the country.

The occasion was an episode of the ABC television series 'The Inventors'. Billed as a launching pad for up-and-coming Edisons, the popular program was closer to a panel show than an analytical forum but it served to publicise many ingenious Aussie inventors.

Compered by Geoff Stone, 'The Inventors' had a panel of judges which included the Lord Mayor of Sydney, Leo Port, and socialite Diana Fisher. After talking to some local engineers and viewing a film of the engine whirring away in Sarich's workshop driving a small generator, the panel was convinced that Australia had a world-beater on its hands. In spite of the fact that no member of the panel had driven a car fitted with a Sarich engine, they crowned Ralph Sarich 'Inventor of the Year'.

In its defence, the TV program was aimed at a mass audience and intended to promote ideas which were still in an infant state and thus help the inventor gain commercial support. The award netted Sarich a few hundred dollars and enormous publicity.

The latter is understandable. In the early 1970s, the engineering world was rotary-mad. It seemed generally accepted that, after nearly 100 years of going up and down, it was time that the internal combustion engine went around and around.

1969 had seen the tentative release onto the Australian market of two rotary-powered cars — the NSU Ro80 and the Mazda R100. A string of major car companies, including GM and Rolls-Royce, was known to have signed up for licensing rights to Dr Wankel's unconventional engine. A plethora of new models was expected to follow.

In 1969, 1970 and 1971 almost every edition of each motoring magazine carried a story on a sensational new rotary engine together with 'inside information' about which car company would be next to launch a Wankel-engined model.

In early 1972, Australia's Modern Motor magazine reported rumours that the aluminium Chevrolet Vega engine was to be GM's last completely new piston engine. Although the magazine stressed the speculative nature of this report, it added that 'background information strongly supports the story'. The magazine also suggested that, by 1978, all GM engines would be rotary. The widespread application of rotary engines by Ford was also said to be imminent.

A few months later Modern Motor went one step further, running a banner on its cover stating that General Motors-Holden's, Australia's largest producer of cars, was 'To Go Rotary'. In its September 1972 issue, Australia's biggest-selling car magazine, Wheels, gave its readers some Sarich information which it later claimed was a world scoop.

The report actually covered two new rotaries: one was the brainchild of Vienna-born professor Franz Huf which 'could emerge as the power plant of the future', the other was the 'Amazing New Aussie Rotary' designed by Ralph Sarich, which 'could be a major breakthrough'.

The great excitement generated by the news was further increased by the name.

'The Aussie design is more than a rotary — it's an orbital design' said the cognoscenti, even if they had little idea what orbital meant in this context.

In a marketing sense, the name was a beauty as it tied in with space travel and the high technology associated with the launch of America's Apollo rockets. With Australia's passion for the underdog, it was an added bonus that Sarich was not old and erudite but a young, self-taught engineer who seemed camera-shy.

Not many people understood how the Orbital engine worked (Sarich released no details until his worldwide patent applications were through), but they knew what it looked like because they had seen a barrage of photographs. Most journalists were in the same category as the general public, but this did not stop them making what, at the time, seemed like monumental claims. Some examples published in 1972 include:

* The Sarich Orbital would cost $60 to build in quantity and was so cheap that it could be replaced rather than repaired.
* Major Australian, American, Japanese and British car firms had expressed interest, along with at least one aircraft manufacturer.
* There were only ten major moving parts which made it even simpler than the Wankel engine. The latter in turn was considerably more simple (in principle) than a conventional engine.
* The Orbital engine was the same shape and size as a spare tyre (approximately 41 cm across and 13 cm wide) and could be adapted as a two-stroke design, four-stroke design or as a diesel engine.
* Being extremely light (about 41 kg) and small, it could be installed under the back seat of a car.
* The engine used very little fuel and produced less air pollution than a conventional unit.
* The power output of 150 kW (200 bhp) at 5000 rpm (from a nominal capacity of 3 litres) was similar to that of the gross output of the then-current Holden 308 (5-litre) V8. Some journals even claimed that 200 bhp was a conservative estimate.

The claims went on and on. The engine was said to be remarkably smooth and silent. Its firing impulses were so regular they produced a continuous drone, making the engine unusually quiet even without a muffler. One report said that frictional losses were negligible and therefore the engine would last about twice as long as a normal power plant. (This claim was based on the relatively slow seal velocity.)

Another interesting claim was that the torque output was so high that the car probably would not need a gearbox. (Sarich had indicated that the usual gearbox could be replaced by an hydraulic device because the engine developed high torque at low rpm).

There were also accurate reports that the Western Australian government had promised financial help. Labor's Premier John Tonkin – and the previous Liberal Premier Sir David Brand (who lost office in February 1971) – had both said their governments were prepared to put up several million dollars to get the Orbital engine into production. Tonkin wanted to see an automobile industry in that State.

'Nothing will shake our confidence in Ralph Sarich', said Mr Tonkin '... all my ministers are working to support him ... we have already inspected several potential sites for a manufacturing facility'.

He went on to say that the government guaranteed favourable shipping to the east coast if the engine was mass-produced in WA. Costing exercises had shown that the freight could be kept down to 40 cents a unit.

In the face of such hype, few were prepared to point out that the Sarich engine was similar to some other concepts and that the term 'orbital' had been used in the late 1960s by the New Zealand company Rotarymotive Developments.

Ralph himself was becoming increasingly concerned about the hype.

'It did us a lot of damage', he said when talking to the author in 1988. 'Most of the reporters were non-technical and thought they were doing the right thing by us in exaggerating. Some got carried away with themselves and said ridiculous things.'

Work on the engine started in an unassuming workshop in Morley (a Perth suburb). This workshop measured 12 metres by 10 metres and contained a lathe and milling machine, a few workbenches, a drawing board and little else except a telephone which, rumours suggested, ran hot day and night with foreign executives desperately bidding for the rights to the amazing engine.

Ralph Sarich had started to design a lightweight engine in late

1969 and got down to serious business in 1970. Acting as designer, draftsman, machinist, fitter-and-turner and office boy, he started work on the first prototype. In less than two years it was ready, having been completely machined on a lathe and milling machine in the tiny Sarich workshop.

The two-cycle design had six combustion chambers and a primitive form of two-stream fuel-injection. Unhappily, it was not successful.

'It ran in spasms of power and was very frustrating', said Sarich later. 'The problem was that gas leaked through the ports into the inlet manifold, setting alight the incoming fuel mixture. The manifold caught fire every time it started up.'

To pay for the continuing research, Sarich was spending the evenings and weekends doing contract work, making or repairing any type of machinery customers wanted to bring him. He also became involved in production work. In one instance, a belt-driven lathe of early twentieth century vintage was converted to automatic operation to produce a special type of bolt. It could spit them out faster than more modern equipment, allowing Sarich to undercut the opposition. This machine became a key player in supporting the research into the engine.

By the time the prototype was ready to fire up, Sarich's staff comprised fitter-and-turner Bruce Fairclough, machinist Colin Pumphries, trainee Ken Johnsen and apprentice Neil Sarich. They all made significant contributions to the company. Neil Sarich left the firm during the early 1980s but Ken Johnsen is currently the executive in charge of the Orbital Engine Company (OEC), Bruce Fairclough is a senior technician and Colin an engineer with OEC.

Deeply disappointed by the apparent failure of his first engine, Sarich accepted advice that the car makers would be more interested in a four-cycle unit. He modified the original design to run as a four-stroke. He replaced the fuel-injection system with a carburettor and made various refinements or changes to the sealing during construction. This engine was fired successfully for the first time on a Sunday morning, 18 June 1972.

It was a time of great celebration. Ralph phoned Pat who rushed around with their young children. He called Ken Johnsen, who was having a morning off for once, and he came over with his father. At the time, the 17-year-old Ken had been with the company for only three months and was surprised at the jubilation.

'I didn't fully appreciate what firing up the engine meant,' he said later. 'I had assumed that when we put it together it would run. I remember that it started easily and was responsive to the throttle. The engine ran into cooling problems after five minutes — but that was long enough to tell us that the concept worked.

'During the next few days it seemed as though the entire world's press had descended upon us. Later we had a visit from a guy called Bob Ward and his boss Dr Hoskins from BHP and I began to realise just what Ralph had achieved.'

The achievement was that Sarich had a completely new engine running just two years after conception.

'I got a terrific kick out of the fact that we had done it a lot quicker than the Germans did with the Wankel rotary', he said later. 'According to reports, it took Felix Wankel at least ten years to successfully operate his engine due to massive problems with the sealing system.'

Sarich's new, four-cycle engine was shown to the press, mounted on soft-rubber mountings in a simple frame and turned by an electric motor in place of a starter motor. In spite of the soft rubber mountings, journalists commented that the engine didn't jump around as expected and was unbelievably smooth. What they - and Sarich — didn't know was exactly how much or how little power it developed.

By this time, Sarich had two partners — Anthony Constantine, a bachelor of mechanical engineering who managed the Wundowie Iron Works near Perth, and Henry Roy Young who ran an earth-moving company at Carnarvon, 11,200 km north of Perth. The three men established a venture which aimed to conduct research and eventually produce engines and they were prepared to spend about $3000 a month to keep the project going.

In June 1972, when Sarich first ran his engine, Middle East crude oil was $US2.18 a barrel and petrol about 11 cents a litre at Australian pumps. The following year brought the first of the OPEC price shocks and increased the demand for fuel-efficient engines. Things could not have looked better for Sarich.

Taking its cue from the renewed interest in fuel economy, the West Australia, a conservative morning paper, said 'Sarich need only ask to get millions'. After the intitial burst of publicity, a batch of six engines was commenced. All were seven-chamber, four-cycle petrol units and none ran on diesel fuel as some later reports claimed.

One newspaper at this time reported that Ford and others had offered cars to serve as test beds and Sarich later confirmed that he had received a free Cortina and a Renault 17 for this purpose. He said that, apart from the cars, he received no assistance from Ford or Renault. Makers of other plane, tractor, boat and other engines also approached him, partly because of a published claim that a senior official from the Department of Civil Aviation had inspected the engine and enthusiastically declared it could be ideal for light aircraft.

During the media hype of mid-1972, Sarich announced he would have a working engine ready for Christmas. At that point, a thorough dynamometer test would be done and an engine would be fitted to a car for long-term testing. The trials would be supervised by a professional engineer, Ian Miller, who was then deputy general manager of the Royal Automobile Club of Western Australia. He was — and still is — held in high regard by Sarich.

Sarich was quoted as saying that the power of his engine could be doubled with a 50 per cent increase in engine weight. That equation — 300 kW (400 bhp) from 68 kg — suggested that he had a world-beating racecar engine as well as one suitable for almost anything else which moved. However Sarich later said he had qualified the original statement, indicating that the internal width of the prototype could be doubled for a 50 per cent increase in overall weight, thus producing a 3.5-litre unit capable of 200 bhp when fully developed.

More published claims came thick and fast. In its latest valveless form (the early four-cycle prototypes had poppet valves but Sarich was developing a disc system which allowed him to dispense with them), the engine would be sold completely sealed. It would need no maintenance or adjustments, but would be replaced at around 320,000 km.

The 22 September edition of the now-defunct Australian Motoring News reported that the most exciting automotive engineering in the world was being done in a small Perth workshop. The engine, the magazine said, 'has been described as the greatest advance since the Otto cycle internal combustion engine'. By whom, it did not say.

'I have seen how the engine operates and listened to it run', wrote the author John Rudd. 'It is almost unbelievably silent and smoother than any conventional design — V or in-line formation, racing or road-going — that I have ever seen.'

This was sensational stuff and the general media took up the call. Here in our own backyard we had a man who could set the automotive world alight and nobody was funding him. Sure, the Western Australian government had made early promises — and Sarich and his partners had put in $120,000 — but millions of dollars were needed to take on the big multinationals, stated one journal.

On 29 November 1972 the call was answered. Australia's largest company, BHP put its name on the dotted line. A company spokesman said it would spend up to $50 million dollars on the project if the engine proved feasible. The second half of this statement was quickly forgotten.

The announcement prompted front-page stories around the nation. Sarich even posed with Miss Australia, Michelle Downes. Together they held up the disc-like engine for the cameras while Sarich told a reporter that the project 'may only cost BHP $100,000 if we can develop it quickly and without problems'. Elsewhere he was quoted as saying that 'if everything goes well, the engine could be ready for production within two years'.

Though a big surprise to many, the BHP deal was no snap decision. BHP representatives had contacted Sarich only hours after he appeared on 'The Inventors' program and the deal had been negotiated in the intervening months. Editorials praised BHP's decision and WA's Premier Tonkin said his government was enthusiastic about the 'marvellous breakthrough' which the BHP offer represented.

Soon afterwards, the Bulletin magazine reported that practically every car maker and many other companies had made offers before Sarich elected to go with BHP. One Japanese company, the magazine said, had reputedly offered $4 million outright for the engine and a US bank was trying to buy a one per cent share for $250,000. BHP could spend a further $200 million on the engine if it proved a goer, the magazine added.

Royal car-buff Anthony Armstrong Jones (Lord Snowden) was one of many who called at Sarich's workshop in late 1972. He inspected the engine at length and offered to drive an Orbital-powered car across Australia to help test it. The offer was not taken up.

In December 1972, the front page of Perth's Sunday Times shouted 'Good news for 1973 — Sarich engine to be built in Perth'. The story said that production would start within 12 months.

Although the car powered by an Orbital engine failed to greet the Yuletide as promised, the enthusiasm continued into 1973. Among others who pronounced on the subject was Ian Miller who was to supervise the testing on behalf of the Royal Automobile Club of Western Australia. He was quoted as calling the Orbital engine 'the most significant breakthrough in technological development — in the automotive industry particularly — this century'.

In January 1973, BHP and Sarich formally set up the Orbital Engine Company as an equally owned entity to 'develop, license, manufacture and market the Orbital engine'. That month Sarich produced more headlines than almost anyone in Australia, seemingly eclipsing even the newly elected Prime Minister, Gough Whitlam, who was the first Labor man to fill that post for 23 years.

Wishful thinking had totally outrun hard cold facts and as one journal produced claims about the miracle engine, the opposition went one better. It became a huge game of leap frog.

'I was often misquoted', said Sarich in 1988, 'because qualifications were rarely if ever applied. And there were instances where statements were fabricated and attributed to me, though I had no knowledge of them. On one occasion, Ian Miller of the Royal Auto Club answered a reporter's questions and his answers were attributed to me.

'It seemed as though each journal wanted to get a better story than another journal and built upon a statement or idea which was wrong in the first place. Many published stories were grossly exaggerated or right out of kilter and the next guy writing about me would read a story which was completely wrong and use it as a springboard for higher things.

'The situation got completely out of control. At one stage I would not talk to the press at all, but this backfired because some reporters starting digging around for old stories and "improved" them to make a new story.'

Sarich says he does not know the source of a lot of the statements that were published. He has no idea why, for example, in January 1973, the entire front page of the Sydney Sunday Telegraph proclaimed: 'Steel Giant Stuns Motor Industry: BHP to build People's Car'.

The story – attributed to Christopher Beck – claimed that BHP was negotiating with a car manufacturer, believed to be Renault Australia Pty Ltd, to jointly build a family-sized, six-seater Sarich-brand car. This car had been 'designed for Mr Sarich by Wayne Draper, 25, of Melbourne', the story added.

The article featured a drawing of Draper's futuristic design and painted the picture of local car makers shaking in their boots at the news. The paper reported that although technical details and dimensions of the new car hadn't been worked out 'it is expected to be inexpensive compared with similar-sized models on the market'.

Many of the claims previously made for the Sarich engine were repeated and the writer threw in an intriguing quote from Draper: 'With the Sarich engine being so small, all you've got to do is fit the passengers in ... we no longer have to design the car around the engine'.

Beck added: 'The new car could also benefit from Sarich's gearbox, described by one car manufacturing executive as "the best we've ever seen". The executive was unnamed.

A curious feature of the Sunday Telegraph article is that the newspaper failed to speak with Sarich who in turn had never spoken to Wayne Draper. Draper was a designer who had done some work for Renault Australia. There was never any arrangement that he was to design Sarich's cars and Sarich later received a letter from Mr Draper apologising for a 'little bit of a mix-up in publicity'. Draper had actually been commissioned by Modern Motor magazine to design a theoretical car around Sarich's engine and the Sunday Telegraph had somehow obtained the information before it appeared in Modern Motor's April 1973 edition.

In the same edition of the Sunday Telegraph, a breakout story by a finance writer stressed that this was Australia's big chance to lead the world by exporting the engine to the Common Market and North America, 'although inevitably the engine would be produced under licence overseas'.

Although the Sunday Telegraph had got many things wrong, there was a genuine Renault/Sarich connection. Renault Australia had offered to make cars available to act as test beds for the engine. Rod Slater, Renault's Perth distributor and a former WA rally champion, sparked Renault Australia's interest in Sarich. Following investigation, Maurice Fertey, Renault's local managing director, highly praised Sarich and said: 'I have every confidence in the engine and his ability to make it work'.

He said that only Renault Australia was linked with Sarich 'at this stage', but the parent company, owned by the French government, was 'highly interested'.

At this time (early 1973), no engine had been fitted to a car or tested independently and none of the journalists making the claims really knew how the Sarich design worked. Some had seen the outside of it and had heard brief descriptions from Sarich who had his patent application pending and was reluctant to release technical details. Sarich also decided to follow the Japanese idea of testing the engine not in a car but on a dynamometer where he could simulate the required road conditions. He said this was a more scientific approach as more accurate data could be obtained.

He also decided to release information only on the displacer (i.e. the hardware in which the fuel was compressed and ignited) but not on the stratified combustion process or the two-stream injection system which could not be fully developed until the engine hardware was reliable. Only a few outsiders (mainly at BHP, the Royal Automobile Club and the University of Melbourne) knew about the combustion process. He was terrified that future patents could be invalidated by prior publication.

On 17 January 1973, specifications for the Sarich engine (but not combustion process) became available for public inspection at the Commonwealth Patents Office. This sparked a massive round of press reports describing in words and diagrams exactly how the miracle was achieved.

It was possibly the only time in history that the Australian tabloids had run detailed technical drawings and a full description of a mechanical device in the early news pages.

The Sydney Sun, which ceased publication in March 1988, described the engine in detail and added (inaccurately) that it could be run as a steam turbine as well as a petrol or diesel unit. Littering its copy with spelling mistakes and literals, as though the news was too hot to be delayed by subbing, the Sun said: 'Mr Sarich says his design has overcome to a great extent one of the greatest problems in rotary engine designs, a high rate of chamber seal were [sic] with its resultant high repair costs'.

Quoting BHP's general manager of research, Dr R. Ward, the newspaper said BHP 'expects to have plans for production and marketing the engine ready by the end of thsi [sic] year'.

Brisbane's Courier Mail responded to the Orbital patent release by running nine diagrams of the engine which had 'captured the world's imagination'. It also featured a technical-sounding description of the combustion process which, to this author at least, made little sense.

In a six-page feature on 20 January headlined 'Into Orbit with Sarich and his Fantastic Engine', the Bulletin magazine unveiled a fresh crop of claims.

Written by Jeremy Webb, the article said in essence:

* Renault was expected to announce shortly the first testing program for the engine.
* For the past year there had been an all-out scramble for 'a piece of the Orbital action'. At least one of Detroit's big three had offered several million dollars for the patent rights and similar offers had come from a number of Japanese and European manufacturers.
* The Orbital engine had 'been the subject of exhaustive examination by some of the world's foremost engineers' and had 'come through well'.
* The engine is 'absurdly simple and cheap to produce'. Latest estimates are $80 per unit under mass-production conditions.
* Hawker de Havilland is soon to get a prototype for testing and current predictions are that the first Orbital-engined aircraft could be airborne by 1975.
* A new power plant was urgently needed. The 1976 emission standards were so extreme that 'many car makers say flatly that unless enormously expensive modifications are carried out, the 1976 emission standards will be impossible for the reciprocating engine to meet'.

If these claims were over-the-top, the writer attempted to vindicate the soundness of the Sarich design with the statement that 'it is just not in the nature of BHP to gamble'. The truth is that BHP, before and since, has indulged in all kinds of speculative ventures, most notably in oil and mineral exploration.

'They didn't back the engine because they liked me', Sarich told the author in 1988. 'It was nothing different from buying an equity in a mining venture. They backed it for good business reasons.'

The Bulletin story stated that the latest prototype was 32 cm wide and 46 cm in diameter and weighed 45 kg. Webb added that Eric Milkins, lecturer in engineering at the University of Melbourne, was assisting Sarich in research and he quoted Milkins as saying: 'The engine represents not just one breakthrough — it bristles with them'.

Another interesting remark was attributed to Geoff Robinson, an airworthiness surveyor with the Department of Civil Aviation: 'I believe the Sarich engine has the potential of becoming the prime mover we have been looking for. Coupled with quiet, vibration-free operation you have an ideal aircraft power plant'.

The Sunday Telegraph had Sarich back on the front page on 21 January, this time with the news that the Sarich engine would soon be on the road. The new Orbital Engines Pty Ltd company was working 24 hours a day producing as many prototypes as possible 'regardless of the cost', the paper said.

In spite of the optimism of the first few paragraphs, some doubts were apparent. Sarich was quoted as saying that the firm had a 60 to 65 per cent chance of success, adding: 'We are designing prototypes to reduce lead pollution to acceptable levels and to increase the engine's efficiency'.

The writer stated that Renault had provided Sarich with facilities on a first 'bite of the cherry' basis, but that industry sources (unnamed) believed the Japanese would be among the first to use the engine.

The Sunday Telegraph went on to say that the latest engines bore little resemblence to the early design which had six chambers and a spark plug for each. 'The present prototypes work on a continuous combustion principle, with fuel burning in a continuous stream', it said.

Sarich later confirmed that he had said this was a possibility. He had worked out an ingenious way to house a small pocket of continuous burning gas in one of the engine's disc valves. It would carry the combustion process from one chamber to the next.

'Fortunately we soon learned not to pursue this route. The idea would have worked but exotic materials would be necessary to handle the intense heat and there would have been tremendous sealing problems,' he told this author in 1988.

The Sydney Morning Herald (23 January 1973) commented that Sarich could become one of the world's richest people. The story 'from a correspondent in Perth' stressed that Sarich is 'careful to point out that his revolutionary engine has not yet been proved'.

Few reporters had picked up on this angle. Those who did buried the thought deep in the story.

The Herald painted the picture of Sarich being courted by BHP and implied that he had declined their successive proposals until he got precisely what he wanted. The final deal gave Sarich a greater financial cut than BHP 'though they have as much say. Since then they have agreed to pay him a handsome salary as well', it added.

'Similarly with the motor companies of the world. One of the big boys from Detroit tried to bulldoze him and was sent off with tail between legs. The English companies were amazed when he rejected their offers.

'What he eventually agreed to do with Renault was that Orbital Engine Company Pty Ltd would continue to be the sole manufacturer, in Perth, of the engines. Neither Renault nor any other company would make the engines under licence.'

Sarich was quoted as saying 'I enjoy engineering much better, but I don't mind the business side' and adding that, in about a year, he would pull out of the company as an active partner and return to research.

Also in January 1973, the Sunday Telegraph returned to its favourite theme and ran a page-five story about the scramble for Sarich rights among the world's car makers (with 'Japanese companies pressing hard for a deal within the next few weeks'). The paper headlined its finance section 'SARICH ENGINE WILL HURT SOME COMPANIES' (by a Staff Reporter). The story warned investors to weigh up the implications of the radical changes which the Sarich engine could bring to the industry.

Amongst several samples of speculative comment, it said: 'In listed companies, The Broken Hill Proprietary Co Ltd would gain handsomely. But the big spare parts maker, Repco Ltd, could be a loser [because of the reliability of the Orbital engine].'

Calculations were then made on royalties, coming to the conclusion that BHP could earn $15 million after tax from licensing the engine, a figure which could raise the share price from $11.65 to $14 or better.

The speculation continued eleswhere. On January 29, the Sydney Morning Herald ran 19 separate diagrams plus an annotated photograph and elevated the Sarich engine to new heights by saying that in its petrol form it was 'almost pollution-free'.

The story continued the often quoted comparison with a V8 engine, saying that the latter had 400 moving parts against Sarich's 9 to 15 moving parts (the number of parts was said to vary depending whether the engine ran as a two-cycle or four-cycle design).

The motor magazines also continued the rotary/orbiting blitzkrieg. The January 1973 edition of Wheels magazine had another 'Wankel-beater', an American engine called the Karol-Ansdale Rotary. And every issue of other Aussie car magazines was littered with comments about which company was doing what with which rotary. Despite the competition from overseas, Sarich continued to be newsworthy, making the cover of Wheels (and a sizeable percentage of the total content) in March, August and October.

The English motor magazines were hot on the scent too. The story had broken in England in mid-1972 and, in February 1973, it was announced 'Disposable Rotary Ready This June'.

It was in fact to be 15 more years before Sarich technology had reached the point where mass production became a real possibility.

3

Doubt creeps in

Like everything else, Sarich's dream run with the press had to end.

The first serious questioning of the project occurred when some local publications and the respected English magazine Motor, dated 24 February 1973, carried a story headlined 'Not-So-Super Sarich'. This story was not, however, inspired by the tall poppy syndrome which would have inevitably turned the tabloids on Sarich at some point. It was an analysis of the Orbital engine by one of Australia's best-known automotive engineers, Phil Irving.

Irving is noted for his design on the Vincent HRD Rapide superbike in the 1940s and later his work on reciprocating automotive engines including the Repco V8 which earned Jack Brabham and Denis Hulme successive Formula One World Championships in 1966 and 1967. Irving, who is also a writer and public speaker of uncommon ability, has a reputation for not tolerating any technical nonsense.

After pointing out that few inventions, if any, had fired the public imagination as much as the Sarich Orbital engine, Irving lambasted what he called the succession of 'unduly optimistic and over-coloured reports, based mainly on wishful thinking disguised under the label "theory" ... which fail to point out that it [the Orbital engine] has many bad features which are going to take a long time to overcome'.

In some places, Irving's article was construed as a direct attack on Sarich, but Irving saw it as a long overdue plea for scientific examination of the wild claims which had been made for the engine in numerous press articles.

Like everyone else, Irving had not had an opportunity to witness the engine under test conditions and had worked from the patent specifications and sketchy material issued by the Sarich office. After explaining the workings of the engine, he described what he saw as the inherent design problems, mentioning poor sealing, combustion inefficiency and heat loss through the piston crowns and vanes. He added that, because of balance problems, the engine would require external balancing weights and that this, together with the need for additional ancillaries, would further cancel weight savings. He also made these points:

* 'Nobody has yet seen an engine working, nevertheless statements that a 2-litre engine will produce 200 bhp (150 kW) and weigh only 100 lbs [45 kg] are repeatedly made and avidly assimilated by non-technical people who are, no doubt, influenced by the BHP Steel Co.'s announcement that they would back the project to the tune of $A50 million, forgetting that this money will be forthcoming only if the engine proves to be worth developing — a very big "if".'
* 'Anyone who is familiar with engine design, manufacture or servicing will be able to detect features in the invention which render the predicted performance highly unlikely, even after years of work, and would also be able to put the finger unerringly on many places where expense or trouble might be expected during manufacture or in service.'
* 'Another rash statement which will not bear examination is that manufacture will be easy because of the small number of parts involved. Of course, this largely depends on what is meant by "parts", but the Sarich [unit] would seem to be considerably more complicated than a conventional engine of the same capacity.'

Irving ended his article by saying: 'Time will tell. Not until a complete engine is built and subjected to dynamometer tests will the real worth or weaknesses of the device be revealed.'

In hindsight, it is interesting to note that BHP did back their judgment with real money — roughly $17 million in research funds — and that the final version of the Orbital engine, built nearly ten years later, developed about 90 kW or 120 bhp nett. This output is closer to the original target than it seems. In the 1970s, most people talked in terms of 'gross power' (i.e. without the exhaust system and some power-consuming ancillary equipment such as the generator and water pump).

By the early 1980s, the industry had switched to 'nett power', which is much closer to the actual power as installed in the car. By the standards of the early 1970s, the Orbital engine was developing about 170 gross horsepower.

Other uncomplimentary articles followed the Irving work, the most common criticism being that the engine had still not been publicly demonstrated in a motor vehicle.

Sarich was angered by them and was especially annoyed by Irving's comments. He said that Irving was 'far from an authority on emission regulations because his experience was emphatically biased towards gas-guzzling racing engines'. He also said that he had given information about the novel combustion process only to 'a few trusted people' and that Irving had not taken into account the two-fluid injection or stratified combustion process which Sarich believed would overcome the efficiency problems. From Irving's viewpoint, Sarich had not, at the time, publicly disclosed this new technology (nor had he made it work properly) and Irving had to make do with the available information.

'I knew from the start that two-fluid injection was the way to go', said Sarich in 1988. 'The idea had been used before but it was achieved using large pumps. We developed a low-cost means of injecting the fuel and have a patent for sending compressed air directly through a reservoir of fluid. We called it a "holding chamber".

'We made our first experiments here in 1971 before we fired the first engine, and although it did not work as well as I had hoped, we knew from the start that this was the way to improved fuel economy and combustion stability.

'The very shape of our combustion chambers provided a type of stratified environment and we were working to take full advantage of it. The Royal Automobile Club published an article about this technique, which we called "two-stage combustion" though it was essentially a stratified process.

'Irving in essence called it a "heap of rubbish" and said that I did not have any understanding of the combustion process. It was a scathing attack. I had not met him nor discussed the process with him. Yet the very thing that he criticised then is making us successful today.

'I must admit that the hype which was heaped on the engine from day one had a lot to do with the subsequent shafting we had from the press. Some articles were so far off the beam that people questioned my credibility, thinking I had made the wild claims.

'I accept full responsibility for letting the situation get out of control. We were naive in those days and did not realise how long the job would take. We had built the first engine so quickly that I believed we would have it powering test cars far more quickly than we did. And it was our failure to produce cars which the media could drive which caused us so much trouble.

'Here I was, a fitter-and-turner with no public relations experience in front of a very powerful media. I thought you could make statements with qualifications and that's how they would be printed – but time and time again the qualification was left out.

'A classic example is that when one reporter asked me about the power output, I said that the engine was designed for and expected to reach 200 horsepower when fully developed. The next thing I knew, there were headlines saying that the engine was producing 200 horsepower and this figure became an accepted "fact" in subsequent articles.

'My credibility suffered, and suffered badly. Even today [1988] some sections of the media refuse to draw a distinction between the early days and what has happened since.

'Yet now, the biggest companies in the world are negotiating with us and Australia's largest company is our business partner. Major Australian and US financial institutions are investing in us.

'Despite this, we get very little credit in some areas.

'Even in recent times, I've been called a conman and worse. Who have I conned? BHP, Ford, the banks who have invested in us? If so, why do they continue to do business with us?'

There can be little doubt that the media itself and Sarich's own lack of experience in dealing with the media created a situation which portrayed him in an increasingly poor light. He could have employed a public relations firm to put out regular reports and deny the wild claims being made and often attributed to him. Instead, all available funds were directed towards research, and information about the engine was allowed to dribble out piecemeal, usually as a result of a quick phone call. Even that information was often 'interpreted' by a young reporter with little technical knowledge.

In the absence of 'official' material, newspapers and magazines continued to pour on the hype. There was an avalanche of publicity when information finally appeared on the engine's operating principles. The March 1973 edition of Wheels magazine became the umpteenth publication to bring 'exclusive first details' of how the engine worked. (In its defence, the magazine may well have beaten others in getting the story to press but the long publication lead-time meant it appeared after a similar report had appeared in several other journals.)

The specifications and diagrams included with this article were held up as definite proof that the device was truly a 'mechanical gem'.

The story also said that Renault France was 'almost certainly' about to sign a contract to buy engines and that Renault would make 100 cars available as mobile test beds when the contract was signed. These cars would be released for consumer testing.

Few journals got more excited than Modern Motor, then under the editorship of Rob Luck. In March 1973, Modern Motor headlined:

'SCOOP – SARICH GEARBOX BREAKTHROUGH'

The magazine had dug up the three-year-old patent for Sarich's gearbox and had a former GM research engineer, Collyn Rivers, explain and evaluate it. This part, however, was used at the end of the article. The first section was written by Rob Luck who praised Sarich's 'deep personal integrity' and labelled him 'the inventive genius'. Among many statements, he wrote:

* 'Sarich has now eroded the last trace of industry doubt.'
* 'A major feature of the Sarich engine is the huge reserves of low-down torque' and 'Yes this engine is capable of developing no less than 200 bhp!' [in common with other journalists, Rob Luck had not tested it, though he had visited Sarich and seen the engine in action].
* The Sarich unit 'is the only petrol burning engine yet devised that has a hope in hell of meeting future anti-pollution legislation and, on that score alone, its success is virtually assured'. (Luck was alluding to several statements by major companies that they could not meet the proposed 1976 exhaust emission levels.)

Having whetted its readers' appetites, Modern Motor crammed as much Sarich material as it could in the next few issues.

The April issue contained colour photos, 'dozens of sensational facts not previously covered', and the news that 'Rob Luck feels he [Sarich] is on the verge of making an important announcement connected with Ford and future development'.

The accolade 'genius' was liberally sprinkled through the copy and this gem of prose deserves a place in motor journalism's hall of fame:

'When Cyclone Sarich first stirred and teased at the westernmost edge of our continent more than six months ago, he toyed with the emotions of a continent that was hungry for something totally original that could weld them into something quite patriotic and unified ... able to repel the advances of the strongest foreign invader. He demonstrated his power and resilience by resisting the advances of the world's most powerful corporate raiders. The aftermath was a fever touched with jingoism that ran rampant in the West and gradually spread Eastwards, infecting the majority with an epidemic patriotism most had forgotten or failed to recognise they were capable of.'

The interview which followed could only be an anticlimax. The long piece concentrated on generalities, with Sarich talking about generous assistance from Renault and Ford and how he now had all the best workshop equipment thanks to BHP. He had increased his staff from six to ten and was about to look for more engineers. He talked about air-conditioning, alternative power sources and other subjects but was not subjected to critical questioning.

Luck quoted Sarich as saying Lucas and Bosch were helping on the electrical side and that he was amazed how eager Comalco and several other companies had been to help and how convinced they are of its success. Buried deep in the copy was Sarich's forthright reservation: 'I feel we've got a fair way to go yet to prove the engine's got something'.

The magazine returned to the theme of the 'huge offers' that the inventor had turned down. Interestingly, Sarich said his motivation for refusing to let anyone take over his engine was his reading of the Wankel affair in which so many people took over the development that it became fragmented.

In a contrast to the hype adopted by the magazine, Sarich himself came across as realistic and knowledgeable, quite unlike the media picture which had emerged. He made it clear he was entirely aware of the possibility of more development and/or manufacturing problems and admitted that some of his earlier statements, particularly those relating to the proximity of actual on-road tests, were optimistic. He personally accepted a lot of the blame for the hype, saying he had been pretty sloppy with the way he had presented information.

The second part of this interview, which appeared in the following issue of the magazine, consisted of more sections of prose in which Sarich talked about anything and everything. Amid the small talk, it was explained that the Mark IV prototype was about to undergo extensive dynamometer testing. Quotes about the engine from a variety of people were given, including an enthusiastic Dr Bob Ward (BHP) who said: 'We've had every kind of professional engineering assessment on this engine that you could ask for. Everything we know tells us we should go ahead.'

The opinions of all sorts of people were sought for that issue of the magazine, including aviation experts, people from the fuel industry, even tractor makers. Some of the experts interviewed anticipated development problems and commented that people were expecting too much too soon. Luck dismissed Phil Irving's criticism — which had appeared in the rival press — as follows: 'Unfortunately the article lost a great deal of credibility when Irving made some sweeping statements that were variously incorrect from a project viewpoint or mechanically inaccurate.'

This comment inspired a letter from Irving which was pointed and witty. It observed that Modern Motor had virtually become the 'Sarich Gazette' and had denigrated 'my professional integrity, both as engineer and journalist ... and my considered judgement based on 50 years of engine design and construction'.

'In a recent TV interview, Mr Sarich made the surprising statement that I was "all tangled up" because the design I had criticised was not the design that would be made! The new design seemingly is not to be a four-stroke but will utilize "continuous combustion" and have a disc valve, in place of poppets, but no other details were disclosed. If this statement is correct, then the press and public have been drooling over something which is already obsolete and earlier technical critiques become pointless. Now nobody jumps off a tram after he has struggled to catch it, unless it turns out to be the wrong tram, so what are we to believe? Is the widely acclaimed design a flop and, if so, will the next one be any better?

'A curious feature of the situation is that the company most involved in the public mind has maintained a silence compared to which a Trappist monastery would sound like a football final. Perhaps someone in BHP could be persuaded to issue some authoritative comment on the matter.'

To his credit, Luck published Irving's letter but it did not stop his magazine from following with a special report headed 'from Rob Luck, in touch with our new automotive capital of Australia'. Here he revealed that Sarich was successfully testing the engine and had started work on a new and more exciting version.

'The new engine, for which most of the components have already been built, will be a tiny 10 inches [25 cm] in diameter ... yet it will produce the same projected horsepower [as the Mark IV] — around 200 bhp.

'The Perth genius', Luck added, 'continues to defy his critics with his remarkably rapid development progress and his latest test results represent a kind of egg-smearing attack on the red faces of the industry skeptics — what few there are left'.

While stories continued to outrival each other, Sarich and his team worked on.

The Mark IV was largely a redesigned version of Sarich's previous unit and replaced the experimental disc valves with conventional poppet valves. This was the unit Sarich had designed specifically to install in a car for on-road testing. Like the previous unit, it had seven chambers but there was a much-improved and simplified sealing design and a Holden carburettor.

The Mark IV had fired successfully but a few hours after champagne had been drunk in celebration, the engine overheated in the testbed and became damaged. The problem was a fault in the metal-hardening. The unit was stripped, improved (especially in the seals) and modified to make it idle more smoothly. This version became the Mark IVA and proved to be the most successful of the series. Equipped with a bigger radiator, it was initally tested at low loads by deliberately 'choking' the fuel intake. In this 'choked' condition, it produced 30 kW (40 bhp) at 3000 rpm and Sarich said that when the output reached 90 to 97 kW (120-130 bhp), he would install the engine in a car.

Shortly after this, Sarich and his staff — which had grown to number 16 — moved into a new and larger factory close to the old premises at

Morley. Here they had 505 square metres of floor space and room for additional equipment, including a dynamometer.

This writer first met Sarich in early September 1973. Like Irving, I had examined diagrams of the engine from an engineering point of view and, although impressed by the original thinking which had gone into the concept, was sceptical of many of the claims. It was not until 15 years later that I learned from Sarich that he had been testing his two-stream fuel-injection system from the day the first Orbital engine had been built.

In 1973, however, restricted to using the then available information, I wrote a syndicated newspaper report which summed up the state of play but did little to quell the wild speculation and excitement:

'In an unusually frank interview Perth inventor Ralph Sarich told this writer', the report said 'that he is still not certain his Orbital engine can be developed for production and that it would not be competitive with a conventional engine if it were not for its compact size and weight.

'He also said that the Mark V prototype is now being bench-tested before being installed in a Cortina or Renault for on-the-road tests. These should take place by the end of October [1973].

'The new prototype is 17 inches (43 cm) in diameter and 12 inches (30.5 cm) long, including the carburettor. When placed in the engine compartment of a Cortina, it left enough room for two men to stand beside it, he said.

'The prototype has a capacity of 3500 ccs and has been tested under load for periods up to 13½ hours. In these tests, it developed 100 foot-pounds torque at 3000 rpm and has reached speeds of 4500 rpm. Though Mr Sarich would not give horsepower details, these figures suggest a current output of around 100 bhp. The designer claims that the engine will eventually produce somewhere between 130 and 200 horsepower.

'Mr Sarich said that no exhaust emissions or fuel consumption checks have been made but that the engine is currently using more fuel than a conventional design. The current weight is 140 pounds (64 kg), including flywheel and carburettor but not the starter motor or generator. This figure could be reduced to about 100 pounds (45 kg) in a production version.

'On this basis', my report continued, 'the engine would develop twice as much power for its weight as a Wankel rotary and four times as much as a conventional engine.

'Mr Sarich agreed that the design has several drawbacks compared with existing car engines and that fuel consumption is likely to be a problem. However, he says that the engine's extremely compact size makes it ideal for light aircraft, marine and for portable power generators.

'The Federal Government has shown considerable interest in the project and the Minister for External Trade, Dr Cairns, has personally inspected the unit.

'The Orbital Engine Company, which is half owned by BHP, now has a staff of 22 people, including nine on the technical side. The project has cost $150,000 to date and BHP are prepared to continue backing it as long as the engine has a reasonable chance of commercial success.

'Four engines are now being built. If tests continue to be satisfactory, 50 will be built in 1974 for large-scale testing in Australia. The prototype will be coupled to a conventional Borg Warner automatic when tested in a car.'

By October 1973, photos were issued of a Cortina with the Sarich Mark IVA in the engine bay. The car had not been driven by this stage but Sarich had installed the engine for preliminary testing while finishing the batch of Mark V engines, which were to be fitted with disc valves.

Sarich commented that the torque output followed a relatively flat curve and that spare parts supplies were being built up so that more testing could be done. He declined to set a timetable for production.

Nonetheless, this inspired more headlines. One announced 'Sarich Ready For Road Tests' and claimed Sarich planned to have 50 Orbital-engined cars on the roads in the hands of consumers for full-scale testing 'within a few months' and said that Sarich believed an engine could be in mass production within two years.

In complete contrast, Dr Ward, the Chairman of the Orbital Engine Company, warned of possible setbacks and said it could be 1980 before the engine would be ready for production cars.

'We didn't understand the magnitude of the task when we started', he said frankly.

'We are a bit like the people who were playing around with the piston engines 50 years ago – we've still got a large amount to learn.'

Interestingly, he listed one possible stumbling block: the possibility that someone could come up with a better engine in the meantime.

'Road-testing is unlikely to happen before the end of the year', said Dr Ward.

Despite this, and other reservations from the Sarich camp, some newspapers continued to refer to Sarich as 'the man destined to become as famous to motoring as Henry Ford'.

Sarich meanwhile had publicly announced that he hoped to ease himself out of the Orbital Engine Company by the end of the year. He wanted to return to his automatic transmission and carry out research into solar energy for use on farms.

A report in Signature (the Australian Diner's Club magazine) in the last quarter of 1974 quoted Sarich as saying that somewhere in the world (it did not say where) designers were working on a car to be built around his engine. It would be a small, light vehicle with the emphasis on safety features, and at this stage (according to the magazine), 'it looks as if the whole of the front of the car will be a safety barrier'.

By late 1974, the staff at the Orbital Engine Company numbered 33. The total amount so far spent was nearing $500,000, but in spite of the early optimism, the Orbital engine was still a long way from the production line. No cars were released for consumer testing and, for that matter, no journalists were invited to sit behind the wheel of the Orbital-powered Cortina.

The public remained largely unaware of what Sarich and his team had already learned: it takes years, even decades, to fully develop an automotive engine. Possibly this was the reason a whispering campaign gained currency that the Sarich engine was a fizzer.

Once unleashed, the rumour went around the traps with the same speed and intensity as the original report that Australia had unearthed a world-beater.

4

All goes quiet on the western front

In view of the enormous publicity, most observers considered it inevitable that a major manufacturer would have signed up to produce the Orbital engine before the end of 1972.

There had been stories published that US, Japanese and European auto executives were almost begging for the rights yet, at year's end, not one car company had put its name on the dotted line. The wait continued through 1973 and that year also closed without a major announcement.

The delays sparked off a fresh batch of rumours. Some members of an impatient public found comfort in the evergreen myth that an oil company had bought the patents and 'buried the engine' because it used too little fuel. Another furphy was that a major Detroit company had snapped up the Orbital to make sure it did not see the light of day and thus force the car maker to scrap its existing engine factories.

People who believe this kind of story do not understand the fundamental nature of patents and how difficult it is to suppress a sound idea with commercial potential. There was certainly no truth in the rumours that the Orbital design had been buried.

Early in 1974, the first Orbital licensing agreement was signed but the occasion lacked the bells-and-whistles impact the public had anticipated. The agreement gave the Australian company Victa the rights to manufacture a mini Sarich engine for use in lawn-mowers. The Victa company started work on a disc-valve 209 mL (209 ccs) aircooled unit with three chambers. At times the engine looked promising, but after investing considerable time and money, Victa bailed out of the project in mid-1976. No Orbital mowers were sold, though some prototypes were tested.

Victa's Ross Phelps, who was heavily involved in the research, later said: 'As with other rotaries, a major complication was the complexity of sealing the engine. This increased the cost of manufacturing and servicing it and we decided it would not be practical for our application.'

He added that the company had learned a lot and the exercise was not a wasted effort.

Little was heard of Sarich during the second half of the 1970s. Some media reports appeared but they were often tinged with doubt. A few were openly hostile, implying that the whole scheme had been a fraud. Others were more gentle in their attack. 'If this man is going to change the world, he's certainly taking his time', was the general theme.

But Sarich had learned that premature publicity was bad publicity and he started to handle his media contacts with kid gloves. He and his staff continued to work very long hours with dogged determination, constantly curing one problem only to find another. The problems were compounded by the energy crisis which had provoked massive spending by car companies as they sought to improve the conventional engine. Sarich had to constantly revise his targets. He ceased to give interviews to every press reporter seeking one, partly because he had reluctantly concluded that he couldn't overtake 80 years of piston engine development as quickly as he had hoped.

Later he admitted:

'I can't deny I was pretty naive at the time [during the early 1970s]. I had got this motor from paper to test bench so fast that I thought that everything else would move fast too. I was setting records for getting a radical idea to work ... everything else should have been easy. I had no idea of the amount of development involved and the colossal amount of time it takes to train people.

'The Wankel rotary engine was first thought of in 1924 and it wasn't until 1950 they had it running. Even after that, more than a billion dollars was spent developing it. I had a working prototype in a matter of two years and I didn't see why everything else shouldn't move just as fast.'

With the 'backyard' approach now well and truly behind him, Sarich was building up an first-rate team of highly educated engineers and turning the Orbital Engine Company into a world-class research and development facility. There were no headlines in that accomplishment, however.

It was the combination of Sarich's inventive mind, dedication and original approach allied to his workforce's engineering expertise which, nearly two decades after day one, set the automotive world on its ear.

Sarich later conceded that the 1970s had been very difficult and stressful but that he had received tremendous support from his family, his staff and BHP.

'The public will never understand the magnitude of the problems we had during the 1970s', he told the author in 1988.

'They were not all technical difficulties. I can't disclose some because they would reveal the dreadful activities of certain people. I don't owe it to them to keep quiet but I can't see the point of stirring up a controversy which will do no one any good.

'There is absolutely no doubt that some people set out to deliberately undermine us. There were businessmen and politicians in bed together. It may sound like a fairytale – but it is absolute fact. Even later, when we signed a contract with Ford, one financial journal published a false report stating we did not have a licensing agreement but a development agreement. We were asked by the [Perth] Stock Exchange to explain the status of the agreement and this meant there was public doubt about whether we had told the truth. Events proved us right.

'The thing which kept me going throughout the 1970s was my belief that the world needed a compact, lightweight engine.

'I took comfort in the fact that GM had faith in the Wankel and was prepared to pour money into developing it, regardless of the technical problems involved.

'I was confident that if I could solve my technical problems, I had a more attractive package than the Wankel, an engine with a much higher power-to-weight ratio.

'I never lost my faith in the two-stream injection system and stratified combustion either and that faith was eventually justified.'

Pat Sarich agrees that the 1970s were a constant battle for her husband when every success was greeted by a new problem but she says that Ralph never faltered.

'I knew that if I complained about the long hours he worked and the fact that the children seldom saw him, he would have given it all away', she said.

'But I held my tongue because I thought he would never be happy until he knew whether his dream was attainable or not.

'He is a very determined man and anything he says he is going to do, he does. That's what kept driving him, not the idea of making money.

'I felt that as long as he thought the engine would succeed, I should go along with him.

'At the time, the engine was everything to him. He's got a good personality in that nothing really gets him down and he seems able to ride over any setbacks. He handles everything so well, never gets stressed out. At times during the 1970s when the engine did not seem to be getting far, I sometimes wondered how he coped but he did without letting it get him down.'

BHP also refused to lose faith. The 'Big Australian' continued to back the development work to a publicly stated limit of $20 million.

'BHP was excellent throughout those trying years', said Ralph Sarich in 1988. 'We've had our differences from time to time but, overall, the relationship has been good and I don't think BHP was ever given full credit for the risk it was taking.'

'Their involvement was financial not technical, but if they had not backed me, I doubt very much if we would be here today.'

From Sarich's point of view, the one good thing to come out of late 1970s was that the media largely ignored him. Apart from a few 'whatever happened to Sarich?' stories, he was left in peace.

However, the media was soon to rediscover the subject in a quite unexpected way. In June 1980, a strange-looking three-wheeled car with Orbital on the side grabbed the headlines. Competing in the first annual Shell Mileage Marathon at the Warwick Farm race circuit, near Sydney, this vehicle was driven by the company's petite receptionist, Carol Darwin. The ultralight body was so low that Carol had to be suspended by straps – but the car consumed fuel at such a miserly rate that it set a world economy rate of 2684.7 miles per gallon, an extraordinary advance on the previous world fuel economy record of 1684 mpg. (At the time Shell Marathon results were given in imperial measurements.)

The design was the handiwork of Kim Schlunke (later Sarich's chief engineer) and his fellow OEC employees. Ralph was not personally involved in the project but agreed to have the Orbital name attached (and pay some of the expenses) when he realised how dedicated his people had become to the unusual project. All work on the record-breaking machine was done out of company time.

The Marathon requirements stipulated that each competing vehicle must complete ten laps of the race circuit at an average speed of 25 km/h. Of the 25 vehicles which took part, three beat the existing world record, including one driven by ex-Formula One driver and engineer Larry Perkins.

The new record was a sensational engineering achievement but ironically the power plant was not an Orbital design. It was a conventional 10 mL (10 ccs), four-cycle piston engine taken from a model aircraft which had been converted from glow plug to spark plug ignition. It also had a new combustion chamber designed by OEC engineers.

After this victory, the record-breaker, which was dubbed by the media the 'Sarich' car, was paraded through the streets of Perth with a police escort and plenty of fanfare. The then OEC spokesman Ken Johnsen (later company manager) said the team expected to break the 3000 mpg barrier in 1981 and, with a number of minor modifications, hoped eventually to achieve 4000 miles to the gallon.

Although not used in the economy run, the Orbital engine concept was far from finished. Since 1978, OEC had been concentrating on a 3.5-litre, seven-chamber Orbital design. By the early 1980s, there were still some problems to beat, but the engine was developing more power and showing better fuel economy than some conventional engines of similar capacity and it had a big weight advantage.

In one of his rare interviews at that time, Sarich commented 'We ultimately expect the [efficiency] margin over a reciprocating motor to rise'. An equally important asset was that the unconventional engine occupied one-third the space needed for an orthodox engine of similar power.

In mid-1981, Sarich packed some suitcases with data and demonstration films and left on an international tour, accompanied by senior OEC and BHP executives. They visited 16 of the world's top car makers in Europe, Japan and the USA and, on their return, Sarich confidently faced the Australian press. He said that nine companies had 'committed themselves in writing' to going ahead with an evaluation of the engine.

Declaring that 'the worst problems are behind us', Sarich added 'we haven't got all the bugs out but we have achieved most of the objectives we set out to achieve'.

In an effort to get rid of these bugs, the 3.5-litre seven-chamber design gave way to a 2-litre, five-chamber design claimed to be half the weight of a conventional 2-litre engine. Sarich later returned to the seven-chamber design.

By then, five different prototype Orbital engines had been designed and about 20 engines actually made. Sarich is not able to say exactly how many Orbital engines were built because several were modified versions of their predecessors. Components, especially the outer housing, would be taken from a worn or faulty unit and fitted to a new engine. A variety of head designs were used, one called the semi-hemi, another had a quad valve layout. In the quest for greater efficiency, the design was becoming increasingly more complex, thus moving away from the original concept of a simple, low-cost unit.

In 1977, Sarich returned to his original 'two stage' combustion process using compressed air to inject fuel into the combustion chamber. Initially known as 'OSCAR', for Orbital Staged Combustion, it was designed to keep fuel mist in the centre of the combustion chamber to promote the rapid burning of an area of leaner fuel mixture further from the spark plug.

The idea was techically brilliant and improved the Orbital engine beyond their dreams. Better still, tests showed that the concept worked equally well on a reciprocating engine and this opened up an entirely new avenue of research. A number of major car companies became interested and Sarich was able to command large fees selling experimental reciprocating engines to car companies wishing to evaluate the process.

The research bills kept mounting. OEC spent $2 million in the financial year 1979-80 and $3 million in 1980-81. In 1981, BHP announced it would pump a further $3 million into the project. OEC also received some government grants when the government realised the significance of the new fuel-injection system.

When tried in a conventional piston engine, the air-operated injection system produced a fuel savings of 15 per cent and simultaneously reduced the exhaust emission levels. By now, OEC had sophisticated testing equipment and its engineers could demonstrate that a reciprocating engine fitted with the Orbital injection system bettered the requirements of the stringent ADR27 Australian emission laws. It produced one-seventh the normal NOx emissions.

Other advantages were that the inexpensive injection system could be largely made from plastic and was easy to maintain. It was also very durable and tolerant of dirt. By now Sarich was working in ultra-modern facilities in the Perth industrial suburb of Balcatta, 15 minutes north of the city centre with a staff of 60. The company's status as a first-rate engine research facility was recognised when the Federal government engaged it to examine the operation of a stratified charge in a conventional engine with the aim of saving fuel.

Work continued on the Orbital design. In late 1981, Sarich told motoring journalist Rod Easdown that it was not enough to have an engine that was a bit better than the conventional design. It had to be a great deal better if he was to convince companies to change to the Orbital system. Being better was getting harder by the day.

'It must be borne in mind', said Sarich, 'that all the time we were working towards these goals the oil situation was causing the automobile industry to put phenomenal investment into the conventional piston engine to make it run more efficiently. We had to do more than keep pace with their improvements, we had to do better.'

In early 1983, after years of negotiation, GM became the first and only car company to sign an agreement to develop the Orbital engine. GM agreed to pay OEC $100,000 to install one in a car but, ironically, Sarich did not proceed. He and the board of directors had decided to concentrate research on reciprocating engines using the newly developed two-stream injection system and stratified combustion technology.

In November 1983 Dr Bob Ward, BHP's General Manager of Research and Technology, announced that work on the Orbital engine had ceased. The OEC facility would concentrate on developing the fuel-injection system fitted to a two-stroke engine, he said.

'There is a public misconception – a very bad misconception – that we abandoned work on the original engine because we could not reach the required performance levels', said Sarich in 1988.

'That's rubbish. After eight years of hard work, we had the engine properly sealed and the combustion was right too. We were achieving quite remarkable results – they are the words of a GM engineer, not mine. However, I could see that car makers would be reluctant to mass-produce such an unorthodox engine because they would need to completely retool their factories.

'I knew that if we transferred the combustion technology and fuel systems to a conventional engine, we would have the lightweight, fuel-efficient engine I had originally envisaged. And we would have had far quicker returns on our investment.'

His workforce fully understood the reasoning but one very senior man confided to this writer: 'Quitting work on the original engine was the hardest decision Ralph ever made'.

5

The switch to piston power

If necessity is the mother of invention, the Orbital injection system was born one harassing day when Ralph Sarich realised that the Orbital engine's combustion chambers were the wrong shape to promote rapid and complete burning of the fuel mixture.

As there was little he could do to change the shape of the chambers, he set about trying to improve the composition of the fuel mixture itself so that it would burn more quickly and efficiently.

The full dynamics of the technology involved is given in a later chapter. Here it is enough to say that the combustion system is based on the use of compressed air to carry the fuel directly into the combustion chamber. It became known as a two-stream injection system.

The idea itself was not new. Rudolf Diesel had experimented with 'pneumatic injection' during the 1890s but failed to make the system work well. He was forced to use a large and heavy pump to force the mixture into the combustion chambers where a considerable pressure already existed. In the intervening years, several others have tried to develop fuel-injection based on compressed air — but no one solved the considerable technical problems involved.

Sarich recognised that the most efficient form of combustion results from the controlled injection of fuel straight into the combustion chamber (a system used by all diesel engines). He started his injection experiments even before the first Orbital engine was completed in 1971. On the first engine ever built, he used exhaust gas from an adjoining chamber to force fuel directly into the combustion chamber. When this idea was less successful than he had hoped, he used a pump taken from a diesel engine. When that did not work to his satisfaction, he tried injecting the fuel into the inlet manifold where there is negative pressure. Some big technical problems remained, so he switched to a normal carburettor. He felt that research into the injection process at that time would delay work on the engine concept itself.

The idea was put in abeyance and, in 1977, when the Orbital engine was running well, he decided to further investigate the injection system to improve the fuel economy and reduce the exhaust emissions. The actual hardware was developed by OEC engineers Mike Mckay and Ken Johnsen, working under Ralph's direction.

The Sarich system differs from previous attempts. Rudolf Diesel and other inventors had tried to carry the fuel into the engine by blowing compressed air over a reservoir of fuel. Sarich hit upon the novel idea of holding fuel captive in a small chamber and blowing air straight through. This caused the air to shear through the fuel and carry it, in highly atomised form, through an orifice at the base.

The system was tried on the Orbital engine and worked very well, adding a new dimension to the stratified chamber combustion already in use (see chapter 12). The company had considerable success with the E-series Orbital engine (built in 1980) using two-stream injection. Work had started on an improved G-series version when Sarich decided to put all of OEC's energy into developing the two-stream injection system for use in conventional piston engines.

Both direct injection and manifold injection systems were tested at OEC and it became increasingly obvious that the system held tremendous potential to reduce the fuel consumption and exhaust emissions of conventional piston engines.

Prototype two-stream manifold injection engines were sold to a major Italian car maker, three Japanese firms, two US companies and two Australian firms. OEC also sold an injection system to a British carburettor firm and granted a licence for manifold injection to James N. Kirby, a Sydney-based engineering firm. Kirby later discontinued work because, the official statement said, they had only two potential customers — Ford Australia and General Motors-Holden's and both were committed to electronic fuel-injection for future engines.

Orbital Engine Company fitted a manifold injection system to a Ford

Escort engine around 1981. Later, direct injection was tried on the same engine. Several Holden Camira (a GM J-car) engines were tested with both forms of injection between 1982 and 1985 and the car was extensively road-tested. OEC then tried the system on a two-cycle engine using direct injection and demonstrated a fuel economy advantage around 30 per cent. At the same time, the exhaust emissions comfortably met Australian legislated requirements.

For some months, Sarich and his team developed manifold and direct injection systems in parallel but it became obvious that the latter offered the greater potential.

'We saw some output and fuel economy advantages with manifold injection', said Kim Schlunke later, 'but they were small when measured against the advantages available from two-cycle direct injection. The system is still available to customers who may want to develop it. It would, for example, be suitable for use with engines burning alcohol fuel.

'There's no doubt though that our greatest technical achievement is the use of a two-phase injector to inject fuel directly into the combustion chamber. This process in turn is seen to greatest advantage in a two-cycle engine.'

The two-cycle design offers several advantages including lower production cost and inherent superior engine balance. But there's another reason OEC decided to concentrate on a two-cycle engine. This type of unit is basically more efficient than a four-cycle unit (for reasons explained in a later chapter). Its main drawbacks, which relate to the fuel economy and emissions level, are eliminated by the two-stream injection without compromising the virtues. OEC engineers made further gains in reducing the emissions by creating a new kind of variable exhaust port and developing a lubrication system to avoid having to mix oil with the fuel.

As designed for the Orbital engine (and later refined for reciprocating units), the process consists of a small engine-driven air pump and a specially designed fuel-injector. The combustion chamber was shaped to ensure maximum turbulence of the mixture once inside the cylinder.

The OEC-designed injector delivers extremely small droplet sizes (about one-fifth the usual diameter) using air pressure at around 5 bar, or 500 kPa. The tiny droplet size means the fuel arrives in a finely atomised form, enhancing the speed with which it mixes with the air and burns. This of course greatly improves the engine's efficiency.

Early in the program, OEC engineers realised that a three-cylinder, two-cycle engine would maximise the advantages offered by the injection system. It would provide substantial cost, packaging, weight and size advantages compared with a conventional four-cylinder engine of similar power.

They decided to use three and not four cylinders because the former allows an optimum exhaust configuration with considerable technical advantages.

As explained in more detail in a later chapter, the exhaust system of a three-cylinder engine allows the exhaust pulse generated by one cylinder to help charge the adjacent cylinder. This effect occurs right across the speed range — making the three-cylinder, two-cycle engine an ideal set-up to take advantage of the injection system.

Engine vibration did not prove a problem, even during early trials, but balance shafts and special engine mounts could be used if a car company wanted to improve the perception of engine smoothness.

Once they had the injection system working well, OEC engineers developed a variable valve-timing device which allows the engine to automatically make subtle adjustments to the timing.

This device is driven by a small DC motor controlled by the engine's management system and acts as a variable height exhaust port. Unlike a similar system developed by Yamaha as a power valve, the Orbital Combustion Process (OCP) unit is designed to improve the emission levels. However, it also contributes to the very flat torque curve, especially at low speeds.

An indirect result of the device is that cars fitted with the engine do not misfire (that characteristic two-stroke, corn-popping sound associated with motorcycles) when a driver takes the foot off the accelerator pedal.

The first three-cylinder, two-cycle engine was based on a Suzuki marine engine as a temporary expedient. At least 20 such units (known as C-types) were built and some were taken apart and rebuilt, bringing the total closer to 30 units. Though very useful for developing the concept, they were difficult to fit into a car because the cylinder block had been designed specifically for marine use. One problem was that the inlet system was on the downside of the engine so, when it was installed in a Holden Camira, a complex manifold was necessary to carry the air from the top of the engine to the underside.

Once the engine was shown to work well, the OEC team designed its own cylinder block and called the new engine the R-series. In this unit, the air was drawn into the crankcase in a manner similar to a modern motorcycle engine. About 30 R-series engines were built. Some were fitted to Chevrolet Spectrums for GM and to Escorts for Ford USA and these were extensively tested by the US customers.

It took about a month to fit the engine in each car and calibrate it for optimum results. Even though the engine behaved spectacularly well compared with the standard unit, even better results could have been achieved by changing the transmission and/or final drive ratio.

The R-series design had some technical limitations because the scavenge axis was at an angle to the crankshaft. This made the engine very long for a given bore size because transfer ports were fitted between each of the bores.

To overcome the drawbacks, OEC designed the current X-series which made its debut in mid-1987. The new unit incorporates its own patented porting system and is designed for high volume production.

Although Sarich has been criticised in some circles for not having the engine in production already, few things move quickly in automotive circles. Despite the uncompromising evidence of dynamometers and on-road tests, there remains an understandable reluctance for car companies to even contemplate switching to two-cycle technology. History is against the two-cycle unit, at least for automotive work, and customers (and rival salesmen) can be expected to belittle the new engine by associating it with the popping, smoky, temperamental units fitted to lawn-mowers and two-wheelers.

The first commercial breakthrough for the Sarich design came in the marine engine field partly because of pending emission legislation but also because most outboard motors are already two-cycle designs. The joint production of Outboard Marine Corporation and Mercury — who have signed licensing agreements — exceeds a million engines per year. Sarich is also negotiating with several Japanese marine engine manufacturers.

For such companies, the major attraction is the engine's low emission levels compared with a conventional two-cycle engine. The better starting, greater idling stability, absence of smoke and improved fuel economy come as a bonus. Motor vehicles are already subject to stringent emission controls and similar legislation is being prepared to include marine engines and other forms of motive power.

The engine offers several additional advantages to car makers.

The low packaging size means that cars could be restyled for better aerodynamics and more usable space. The underbonnet layout can be redesigned to take the spare wheel and other components; the greatly reduced engine weight means that the front structure and suspension can be lightened and tyre wear reduced. The reduced engine length means that wishbone front suspension units are more attractive to use and this complements the low engine height. In some designs at least, there's less need for power steering, giving an additional saving in cost and weight.

From a manufacturing point of view, a crucial advantage is that car companies could use existing tooling to mass-produce the OCP engine, using a reduced number of components. This would lead to a big reduction in manufacturing cost, even after paying a substantial royalty for the technology. One major objection to the Wankel engine, now 'fully sorted' by Mazda, is that costly, specialised equipment is needed to mass-produce it. Even Mazda agrees that, with present volumes, the rotary costs 10 per cent more to produce than a piston engine of comparable output.

Since 1983, when a piston engine using the OCP two-stream injection system was installed in a Holden Camira, considerable progress has been made in refining the concept. Sarich set up his own team to design and make electronic engine management systems and initiatives here have greatly improved the engine's driveability and exhaust emission levels.

The decision to switch from the Orbital design to research into the new fuel-injection system was greeted by some sections of the press as evidence that the original engine had failed. It also caused heartache within OEC.

'Ralph deserves nothing but praise for that decision', a senior OEC officer told the author. 'Some people thought we were throwing the baby out with the bathwater but he knew, and I think we all knew, it was the right thing to do. That didn't make it any easier, though.'

The decision was influenced by several factors.

Apart from the technical attraction of the two-stream injection system, it was becoming increasingly apparent that major car companies were reluctant to depart from the trusty reciprocating engine. At the time only one company — General Motors — was prepared to order an engine for installation in a car, even though Ralph and his sales team had made several world tours. They had developed an impressive presentation demonstrating that the Orbital offered major weight and size advantages and was already as efficient as most piston engines.

However, as quickly as he improved the rotary design, car makers came up with equally significant improvements on the conventional engine. The chase could go on indefinitely and even if Orbital came out in front, there was no guarantee that a major company would take the monumental step of switching to an unconventional power plant.

Sarich had reached the point where a great deal of work was yielding only small gains. The Orbital engine had numerous advantages and there was plenty of scope for further improvements but commercial success — if it came at all — was a long way down the track. His rotary remained in the 'promising' category and, understandably, shareholders wanted something which offered a more immediate return on their investments.

Besides, the new injection system would enable him to crack a knotty problem which had previously defied Ford, General Motors and Texaco — the use of stratified combustion chambers. The concept basically allows the burning of a lean fuel mixture by forming a small volume of rich mixture which burns first and ignites the main charge. Detroit had spent a fortune trying to solve the problems — to no avail. Sarich, on the other hand, had achieved a type of stratified combustion chamber with his Orbital engine from day one and was satisfied he could combine the concept with the injection system to produce a new generation of piston engines.

Experience with direct injection had already shown that a notable reduction in fuel consumption and exhaust emissions could be achieved. The exhaust emission readings obtained so far were remarkably good.

In a conventional engine burning unleaded fuel, an expensive two-stage or three-stage catalytic converter is required to convert the major pollutants to water, carbon dioxide and nitrogen. The Sarich engineers found they had fewer emissions and these could be handled with a low-cost oxidation catalyst with a reduced quantity of precious metals.

By early 1983, OEC had established that healthy commercial prospects existed and orders for prototype four-cycle engines and fuel systems were received from several major manufacturers. The work on these units proceeded but was eclipsed by the extraordinary success being achieved with two-cycle designs.

Piston engines are often divided into two categories — four-cycle (or four-stroke) and two-cycle (or two-stroke). Both have an operating cycle comprising four actions: the fuel mixture enters the combustion chamber; it is compressed; it is ignited, burns and expands rapidly; the burnt gases are expelled.

The fundamental difference between the two types is this: in a four-cycle unit, the four basic actions take place while the crankshaft makes two complete revolutions and the piston makes two movements up and down the cylinder. In a two-cycle design, the same four actions occur while the crankshaft makes one complete revolution and the piston moves once up and down the cylinder.

To achieve its operating cycle, a four-cycle unit needs complex poppet valves (or similar) which control the intake of fuel and the exhaust of spent gases. In modern engines, they are usually operated by an overhead camshaft which is chain-driven or gear-driven from the crankshaft. The two-cycle design has a much more simple system in which inlet and exhaust ports are covered and uncovered by the piston as it moves up and down the cylinder.

The vast majority of cars have four-cycle engines which, because of the more complex valve gear, weigh more and are more expensive to produce. However, traditionally, they have several important advantages — smoother slow running, better fuel economy and reduced exhaust emissions.

Many automotive engineers agree that two-cycle piston engines are technically superior, being smaller, lighter and cheaper to make than a four-cycle unit of comparable power. But there's a big gap between theory and practice. Several major car companies have spent considerable time and money trying to develop two-cycle engines for motor cars but have not been able to overcome the huge problems relating to exhaust emissions, fuel economy and low-speed torque output.

After testing the two-stream injection system, Ralph Sarich decided that it could eliminate the basic drawbacks of the two-cycle engine without affecting the advantages. OEC started to put the major emphasis of research on two-cycle units because the low weight, low manufacturing cost and compact size make it compellingly attractive for mass-produced cars.

The very first tests on the Suzuki-based, three-cylinder, two-cycle engine thrilled the Sarich team. The unit demonstrated greatly superior fuel economy, slightly more power and lower exhaust emissions than a conventional four-cycle unit of similar capacity. Tests showed the technology is suitable for any petrol engine from 50 mL (50 ccs) upwards, whether operating on a four-cycle or two-cycle principle.

The major car companies came around to this point of view.

Although Sarich is reluctant to name customers, reports from Europe strongly suggest that Ford will have a Fiesta or Escort (or equivalent size car) powered by an OCP engine on sale by early 1993 and that a larger sports model with V6 OCP could follow. The Japanese customers could be Suzuki or Toyota. Honda seems an unlikely client, at this stage at least, because of the company's stated policy of developing all its technology in-house.

GM's interest in the project goes back several years and at one time seemed to be on the back-burner. Suddenly, things changed rapidly. Someone in Detroit had a big rethink about the project, possibly after testing the OCP-powered Spectrum in Detroit. By July 1988, GM was sounding highly optimistic. Arvin Mueller, manager of the corporation's Buick-Oldsmobile-Cadillac powertrain activities, told Automotive News (the Detroit-based industry newspaper) that a breakthrough in two-cycle ignition technology 'will turn this industry on its side'.

Mr Mueller added that 'the breakthrough technology probably will come from outside suppliers such as the Orbital Engine Company of Australia'.

He said he could not predict that GM would be first with the new product.

'It's hard to schedule an inspiration, but we're close enough in some applications [to say] you'll see it in a smaller car in a couple of years.'

Things continued to move quickly. In August 1988, GM started advertising for engineers familiar with two-cycle engine technology.

Its advertisements were run in newspapers in several cities in Wisconsin where marine engines are made. One such advertisement appeared in Fond du Lac where Mercury marine engines are made. The local paper, the Reporter, outlined a 'challenging career opportunity' for mechanical engineers in GM's advanced product engineering, powertrain and systems group. The advertisement stressed that, in addition to appropriate degrees, the applicant must have considerable experience in 'two-stroke base engine design, tuning, scavenging and structural analysis'.

Other US journals were equally upbeat. In January 1988, the leading US journal Business Week ran a page-and-a-half article headed 'Two-Stroke Engines have Detroit Buzzing'. It talked about the advantages of the Sarich design and then featured a small US company which has spent one million dollars developing four-, six- and eight-cylinder engines called FITs – an acronym for fuel-injected two-stroke.

Discussing the size of the Sarich organisation and the money it has outlaid on development, the article said: 'Judging by what independent experts say, the money was spent wisely'.

'"His group is doing things with a two-stroke engine that have never been done before", raves Michael J. Boerma, president of Michigan Automotive Research Group, an engine development and consulting firm.'

The interest of the world's press as well as the major car companies is easy to see.

As of late 1988, the 1.2-litre, three-cylinder, two-cycle OCP engine had the following specifications as shown in the Orbital Engine Company's annual report. The claims are shown alongside the average figures for an advanced conventional engine as fitted to a standard small car such as a Ford Escort. Note that exceptional power, weight and emission figures are achieved without compromising the fuel economy.

The results were obtained on an engine dynamometer and later verified under vehicle testing.

	OCP 1.2-LITRE UNIT	STANDARD 1.6-LITRE ENGINE
Output in bhp	103	90
Output in kW	78	67
Weight in kilograms	43	128
Fuel consumption (mpg)		
city cycle	36.8	32.0
(mpg) highway cycle	51.8	48.0
Acceleration		
to 60 mph in secs	11.2	14.8
Emissions		
(grams/mile)		
HC	0.17	0.32
CO	0.12	2.80
NOx	0.24	0.80

Although the weight and efficiency advantages are significant, the Japanese firms who have bought development engines have been more impressed by the low packaging size. Styling has become a major feature of the Japanese competitive equation and the OCP unit allows the stylist greater freedom to explore interesting frontal shapes than any engine currently in production.

'The size aspects of the engine cannot be overstated,' says Kim Schlunke. 'It is treated by our Japanese clients as the most attractive feature they can put a value on.'

6

Driving a Sarich-powered car

Apart from the gauges under the dash panel, it looks like a standard Chevrolet Spectrum, the US equivalent of a Holden Gemini. When the ignition key is turned, the cold engine fires instantly.

It seems to be running a little faster than expected but a glance at the tachometer shows the speed is normal — 700 rpm. Touch the accelerator and the engine runs smoothly and evenly. The noise is a little unusual but you can't quite put a finger on the difference — the air intake or perhaps the exhaust system?

No matter. The important thing is to get the car on the road. This is done in the usual way for a manual car — press the clutch pedal, slip into low gear, squeeze the accelerator pedal and let out the clutch. The Spectrum races off briskly and smoothly.

If you experiment with various accelerator positions, you realise that the engine responds far more quickly than the normal unit. It is, in fact, more responsive than any small engine you can remember. When you press the pedal, the car surges forward as though eager to get into its stride.

Experimenting further, you halt on a level stretch of road, tweak the stopwatch and accelerate as rapidly as possible from rest to 60 mph, a speed equivalent to 100 km/h. (Being a US car, the Spectrum is fitted with a speedometer calibrated in imperial measurements.) The car accelerates evenly and cleanly, without a hint of protest from the engine. You note the stop watch time and, to confirm it, repeat the run in the opposite direction. The average figure is 13.2 seconds which is impressive, because the standard car takes 13.5 seconds and its engine is 25 per cent larger in capacity than the Sarich unit.

You remember another factor. All cars are fitted with gearbox and final drive ratios which best complement the engine's torque characteristics. The Spectrum in question had the standard gear ratios, designed to run with the conventional engine which has different characteristics to the OCP unit. Obviously, ratios more in tune with the prototype engine would further improve the performance.

Impressed, you slow in fourth gear until the car is crawling along with the tacho showing 600 rpm. Flatten the accelerator and it takes off smoothly and evenly.

'Why not try that in fifth?' says engineer Mark Lear who is sitting alongside. That sounds like a silly idea, but I try. Remarkably, the car accelerates crisply and evenly. I comment on the amount of torque at low engine speed and the fact that most four-cylinder engines, of any capacity, would stall or run very roughly if subjected to the same treatment.

Mark Lear, who was then a senior project engineer with OEC, agrees with my comment. He boasts that this is no ordinary engine. It's the company's three-cylinder, two-cycle unit fitted with OEC's two-stream fuel-injection, using compressed air to inject fuel directly into a stratified combustion chamber.

Mark Lear explains that the ratio between the fuel and air varies constantly with different operating conditions, being controlled by the engine management system. At times the engine is running as lean as 36 to 1 (2.5 times leaner than the conventional engine), but there is no combustion instability or poor driveability as a consequence.

Under normal conditions, such as cruising down the freeway, the fuel mixture is probably around 21 to 1, which is still considerably leaner than the stoichiometric 'normal' of 14.7 to 1.

The engine itself is capable of revving up to 10,000 rpm, although there is little advantage in taking it above half this speed, even when trying for maximum acceleration.

The above was written immediately after my first experience driving a Sarich-powered car in 1987. I was later to drive better and more powerful versions – but the initial impression was the most lasting.

The big difference between the OCP design and any other two-cycle engine I've handled is that there's an even flow of power throughout the operating range. As used on motorcycles at least, two-cycle engines tend to be 'peaky', developing little power at low speed then, suddenly, full power arrives almost brutally as the rpm increase.

Vibration is always a subjective judgment in a motor vehicle unless there's an array of scientific gear to measure it. But I was not conscious of the vibration level being different from other small cars I've driven. Engine balance also varies with the engine mounts fitted, and these days automotive engineers have a smorgasbord of mounts from which to select. It is largely a matter of finding the type which best suits the engine under development. Mark Lear commented that if a car maker wanted optimum engine smoothness, a pair of balance shafts could be used, as fitted to engines as diverse as the three-cylinder, four-cycle Daihatsu Charade and the BOC/Holden V6.

The most noticeable feature of driving the OCP Spectrum is the reduced amount of engine braking when travelling downhill or when the driver takes the foot off the accelerator pedal.

This is a characteristic of all two-cycle engines and occurs because this type of design does much less pumping. (Pumping is the work done when inducing the fresh charge and exhausting the burned gas.) The OCP engine contributes much less to deceleration than a conventional four-cycle unit but the driver quickly adjusts to it and ceases to be aware of the difference. The unseen drawback is that this increases brake lining or pad wear — a penalty also associated with automatic transmission.

On the subject of deceleration — I did what any other owner of a vehicle with two-cycle motor would do — waited for the characteristic 'pop pop' when I eased off the accelerator. There was absolutely none. This corn-popping noise — so frequently associated with two-cycle engines — is caused by incomplete burning of the fuel within the combustion chamber and the misfiring of the unburned mixture in the exhaust system.

As the small car hummed along the freeway, I wondered how many major car companies are currently experimenting with their own two-cycle engines, if only to avoid paying a royalty to OEC. Later, I put the question to chief engineer Kim Schlunke. He replied that he had no fears about a rival firm building a better mousetrap.

'We have four years start on anyone', he said. 'We intend to stay that way. We have patent protection and technology protection.'

As fitted to the Chevrolet Spectrum I was driving, the three-cylinder, two-cycle unit developed 63 kW (85 bhp) from a capacity of 1.2 litres (the output was later increased to over 100 bhp, thanks to improvements in the engine management and injection systems). Even though this engine was an early example of the breed, it gave a flatter torque curve than a very good four-cycle production engine of similar capacity. I started to wonder how good the supercharged version will be when fully developed.

The standard OCP engine is so light that it can be lifted by one person and lowered into the engine compartment. Measurements show it is one-third the weight of an equivalent four-cylinder, four-cycle engine and occupies one-sixth the packaging space.

The packaging advantage is obvious when you look into the engine bay. There's room, for example, to stow the spare wheel and house a very large tool kit. The lack of weight means that the manual steering is very light in use – a significant advantage in a family car where power steering adds to the production cost.

I did not have the opportunity to check the Spectrum for fuel economy but have no reason to doubt the company's claims that it uses approximately one-quarter less fuel than the standard engine in the same car. The normal vehicle uses 7 litres per 100 kilometres (40 miles per gallon), the Sarich version 5.5 litres/100 km (51 mpg) when subject to the same US urban driving cycle test.

The most significant advantages concern the untreated exhaust emission levels. In published documents, OEC claims these are five times lower than those produced by any two-cycle engine of similar power ever built. The hydrocarbon emissions are 32 per cent lower than the 1988 US limit, nitrogen oxides are down by 50 per cent and carbon monoxide is cut by 92 per cent, using a simple oxidising catalyst.

Of course, a three-cylinder, two-cycle engine does not have the valve train and associated parts needed in a four-cycle engine. This, and the fact that it has three and not four cylinders, makes the OCP unit about 30 per cent cheaper to make than the production engine fitted to the standard version of the same car. The engine (but not the exhaust note) is quieter because the usual valve clatter is absent and there is less internal friction.

Unlike some radical propositions, including the original Sarich Orbital design, there should be no durability problems with the OCP reciprocating unit. Other than the direct injection and combustion systems, the unit follows established two-cycle principles. The engine in the car I drove has done a gruelling test schedule but the only failure had been the air compressor which has since been redesigned.

The bogey of oil smoke and high oil consumption has been eliminated. The engine runs on regular unleaded fuel with a separate oil supply and does not need oil changes as such. Its replacement cartridge can be replaced in a matter of seconds. Company engineers say it uses one gallon of oil per 7400 miles (one litre per 2600 km) which is similar to the amount of oil required for a four-cycle engine given regular oil changes.

As an operating unit, the OCP two-cycle unit is already successful but will it penetrate the volume car field?

Says Kim Schlunke:

'We have a significant cost advantage – estimated at $300-$500 including flow-on savings – pushing us into the marketplace. We have a lightweight, low volume package and the ability to meet all pending emission and fuel consumption legislation.

'We expect to gain a sizeable share of the market.'

He is probably right but first motorists have to throw away existing perceptions of a two-cycle engine. They will only do so if they drive a car with a Sarich power plant. It's a spectacular unit which runs like a four-cycle engine, develops abundant power and pulls like a locomotive at low rpm.

But people won't know until they try one on the road.

7

Sarich — the man and the inventor

Ralph Tony Sarich was born in Baskerville, near Herne Hill, Western Australia, in December 1938. He was one of five children born to Peter and Ika Sarich who had migrated to Australia from Yugoslavia. After they married and had their first son Donald, the couple decided their home country was not a suitable place to raise children. They contemplated migrating to the USA at first but, after discussions with various officials, came to Australia.

Sarich senior arrived first, paying his own fare and travelling by boat. On reaching Perth in 1926, he took a job chopping trees in Pinjarra, a small town halfway between Perth and Bunbury in Western Australia. To save money, he lived in a tent so he could support his wife and young son back home and save for the future.

After five years, he had accumulated enough to pay the boat fare for his wife and son and to buy a small farm in the Swan Valley where the Sarich family settled. By continued frugal living, they were able to save enough to buy a small orchard. Young Donald was joined by Yubol, Ralph, Nellie and Eric in that order.

To pay off the mortgage, Sarich senior ran three jobs. He was employed in a local dried fruit shed, did logging work when available and ran the orchard. He even started a small vineyard, intending to produce fine red and white wines but the only customer he had was the farmer next door.

'It turned out so badly that it was some years before I was able to enjoy a glass of table wine', says Ralph. 'I stayed with beer.'

Ralph recalls his parents as being extremely honest and hardworking people. His mother – a Catholic – was gentle and deeply religious. To her, the biggest crime her children could commit was dishonesty. She frequently told them that even if they managed to deceive their friends and the police, they could never hide from God, not even behind a concrete wall.

Ralph says his father was a cheerful, good-humoured man, very stern at times but just. He used to get up before the sun and work into the night. Mostly he worked seven days a week but occasionally he took a Sunday off to play bowls.

'He kept working until he was 70 and I guess instilled in all his children the value of hard work. As kids on a farming property, we all had to pull our weight, but there were often unexpected rewards. I remember when dad bought my two elder brothers an ex-army, one-tonne utility so they could get into town at night. Don had just returned from fighting the Japanese in New Guinea.

'The ute was something Dad could not afford at the time.'

His parents continued to work hard all their lives and they invested their savings in more land, at one time, owning five small properties in the Swan Valley area. Here they grew oranges, lemons and ran some sheep.

While living on his parents' orchard, Ralph became increasingly interested in the tractor and other farm machinery and this led to a fascination with anything of a mechanical or scientific nature. Ralph's father had little interest in mechanical things but his mother had a technical turn of mind and once told him that her father had been most skilful with his hands and was good with mechanical things.

Ralph showed great aptitude as a handyman from an early age and was soon in charge of all repairs needed around the farm. At school he was an indifferent student. Knowing that his aptitude was more for mechanical things, his father later insisted that young Ralph do a trade course.

With his sister and brothers, Ralph attended the Upper Swan Primary School and then the Midland High School.

'I was not good at school', Ralph now admits. 'My reports said that I had potential but was not realising it because I was disinterested in schoolwork. I think the main problem was that I was unable to concentrate properly when in class. I was easily distracted and did not study. From an early age, I was very creative in a mechanical sense and wanted to be an inventor. I thought much more about mechanical things than formal lessons and could not see the relationship between subjects such as history and geography and the career I intended to follow.'

'I guess my only complaint about school is that I was never told why it is necessary to learn all those things. These days, children must be well educated to survive in this high-tech world.'

Ralph's younger brother Eric became a star footballer playing for the South Melbourne (Victoria) team and later worked as an ABC sporting commentator in Perth. Yubol became a Telecom technician. Younger sister Nell was the best student among the Sarich children and took a job as a bookkeeper after leaving school.

The family remained close and although Donald died during the early 1980s, the others keep in touch. They showed their faith in Ralph by buying shares in Sarich Technologies Trust when it was launched.

Ralph finished his secondary education at the age of 16 and commenced a five-year apprenticeship as a fitter-and-turner with the Midlands railway workshops and worked there until he was 25 years old. Initially, he attended technical college, taking both day and night classes. While still a teenager and going to technical college, he started a small engineering firm which he ran at weekends. He continued studying technical subjects until the age of 22 but did not matriculate or gain a diploma.

To this day, his sole qualifications are confined to his apprenticeship as a fitter-and-turner. He was, however, awarded an Honorary Degree as Doctor of Science by Murdoch University, Western Australia, in 1987.

Looking back on his schooling and night classes, Sarich says that one of his problems was that he always wanted to do things differently from the way he was taught. Rather than accept the conventional way of going about a job, he would study it and try to find a better method. This unconventional approach did not sit well with his teachers who, at times, told him to forget the questions and get on with doing things the way he had been shown.

'By the time my part-time engineering business was going well', he says, 'I concluded that if I devoted to it the hours I was spending at night school, I would be able to employ people to look after the jobs I could not handle. I could also spend my time being creative in ways which were not taught at school.'

One night, all of Ralph Sarich's plans turned into a nightmare. In 1961, he was 22 years old when he almost died in a car accident when returning home from Fremantle after attending a Christmas party for the Singer office where his girlfriend, Pat Richardson, worked. The hour was very late, Ralph had had a few beers and stopped for some sleep on the way home. Apparently, this was not enough as he later fell asleep at the wheel of his FJ Holden. The car slammed into a telegraph pole. He was taken to Royal Perth hospital with multiple fractures, internal injuries and severe facial damage and, for three days, lay on the danger list. After two months, he was released, but had to return to hospital periodically during the next two years for operations to repair his face.

'It was very, very serious', says Pat who is now married to Ralph.

'I didn't think he was going to live. His brother Eric rushed me to the hospital. The priest was already there and no one thought he would pull through.'

The doctors did such a good job, however, that Ralph emerged from hospital in good physical shape and, these days, there are no visible scars on his face. Although he had been told he would never play sport again, he was back at football training within 18 months.

There was a sequel to the accident which commenced in October 1985. Sir George Bedbrook, senior spinal surgeon with Royal Perth Rehabilitation Hospital, phoned Sarich's secretary and asked for an appointment. It was one of many such calls he had made on Perth industrialists during a long quest to raise private funds to continue research into spinal injuries and damage to the motor skeletal system.

'I first saw Sir George in action when I was in hospital after my road accident,' Sarich told the press a few months later.

'I can't say I knew him but I came to admire the work he and his colleagues were doing with accident victims.'

Sir George did not remember the young man in his orthopaedic ward 24 years earlier, but Ralph Sarich had not forgotten Sir George.

And so it happened, in May 1986, that the Sarich family announced a $2 million initial grant (and a further $1 million subsequently) to establish the Medical Foundation of Western Australia (known to the family as Medwa). Sir George and Ralph Sarich sit on the board and Ralph's daughter, Jennifer, acts as honorary secretary and administrator.

Ralph emerged from the 1961 car crash as a different man, not just in appearance but in outlook.

'My brother Eric and I were pretty wild in those days,' he recalls. 'We used to love having a great time. I realised that it was time I settled down. I had met Pat at a dance when she was about 16 and now, five years down the track, I asked her to be my wife. We were married in 1962 in St Marys Cathedral in Perth.'

Blonde, fair, slim and extraordinarily supportive, Pat became the stabilising element that Ralph's life needed. Thanks to her co-operation and help, he was able to pursue his career with single-minded intensity and develop the technology which may change the course of automotive engineering.

Born of English parents who had settled on a dairy farm about 100 km from Perth, Pat had a similar background and family life to Ralph. Today she is his greatest fan.

'I was immediately attracted to Ralph with his dark hair and dark eyes', she told the author. 'He was happy-go-lucky and had a mischievous way and a cheeky grin. He was a bit of a lad especially when out with Eric.

'To be honest, I thought he was pretty immature when we first met. But I ran into him two years later and he seemed to have grown up a lot. He became my first serious boyfriend.

'He was extremely ambitious, even then. We used to sit and talk for hours about the future. We had both come from large, hardworking families, not poor, but hardly wealthy. We both wanted to get somewhere in life and were prepared to work for it.'

And work they did. The road accident had cost Sarich a great deal of money, both in medical expenses and in lost time from work. For the first 18 months of their marriage, Ralph was in and out of hospital. Although they were virtually broke, he and Pat bought a house and small orchard at the outer Perth suburb of Herne Hill from his father on no-deposit terms. Pat added to the family income by managing the orchard and taking a job as a bookkeeper in the local hardware store, while Ralph worked long hours in his workshop and at his job with the railway workshops.

Sarich's assets include an exceptional level of quick intelligence, a fierce determination to see a project through and a background which covers an array of mechanical jobs. Even when still working at the railway workshops and attending night school, he was designing and building agricultural and industrial equipment as well as repairing and modifying existing machinery.

'We did all kinds of work', he says, 'making agricultural equipment, modifying tractors, making loaders, trailers, spray pumps and sometimes hydraulic pumps'.

'I left the railways when I was 25. Pat encouraged me to leave and get experience in private enterprise. I knew what I wanted to do with my life and felt it was essential to get a wide experience in engineering. I went to work with Thiess Brothers for a while as a fitter-and-turner on the standard gauge railway project and soon became engineer in charge.

'I was earning good money, Pat was working and we began investing in real estate. By the time the standard gauge project had finished, I was ready to go into business full-time on my own account.

'I took the lease on a new BP service station which opened at Midland Junction on the Great Northern Highway and made it a condition of the lease that I could open an engineering workshop alongside the service station.

'We did not make much profit selling petrol but received plenty of engineering work. I also got more and more involved in buying and selling real estate, mainly small rural properties.'

It was at this point in his life that Sarich the inventor started to fully emerge. He designed a large earth-moving scoop and sold a dozen or so to neighbouring businesses. He took the theme a step further and designed a self-tipping trailer based on the scoop. He had worked out the geometry in such a way that loading the trailer energised it and, to untip, the operator just pulled a lever. The load would then tip out and, at the same time, the trailer moved to spread the dirt. Once empty, the trailer bed would click back into position.

'I guess we sold about 20 to 30 of them', says Sarich 'but it was never a main line. We never marketed it, the trailer sold by word of mouth.

'Later, when I started the engine firm, I designed an automatic water sprinkler system which we made at Orbital Developments, a small firm I ran to generate income to help pay for the engine research. Later a major company took a licence on the sprinkler and we agreed they could produce their own version which has since been very successful.'

Sarich's other invention of the time was an automatic transmission operating on hydrostatic lines. He fitted it to the family Falcon and proved that it worked.

'It was very compact but not particularly efficient', he now concedes. 'It was built on crude machinery and suffered from hydraulic leakage but at least it gave an infinite number of ratios and allowed the car to travel at all normal road speeds. It also incorporated a lock-up device to save fuel when a one-to-one drive ratio was reached. This, like my real estate dealings, was just a sideline.

'But I began to have mixed feelings about the way our society operates. I could make more money selling one block of land than by working long hours seven days a week. No matter how hard I worked, we never made as much money in the engineering business as I made gambling on real estate.

'Often we had little actual money invested in the land. We could buy on credit and hold it for just a few months and then sell at a profit — often tax-free. We were really rolling it over.

'I learned that it is easier to make money without working — and concluded that the government was encouraging us all in the wrong direction.

'Meanwhile, I was working very long hours at the BP service station so Pat and I decided we should sell the business and I would take a nice comfortable job somewhere and continue to speculate on land.

'In 1966, I joined West Farmers-Tutt Bryant which involved me in both the technical and sales side of heavy equipment. I started as a sales engineer and soon became manager of one of the divisions. I was transferred to Port Hedland (on the north coast of Western Australia) to look after operations between Carnarvon and the Northern Territory. Here we used to deal with large engines many of which I felt were too big and heavy for the job they were doing.'

During this time Sarich began to appreciate how valuable a lightweight, portable power plant would be to men working on remote construction, mining and farming jobs. He never forgot that point and, later, his search for such an engine provided the technology which has brought Sarich worldwide acclaim.

'After two-and-a-half years with the firm, I decided to get into the design business myself.'

Cars have interested Sarich from his earliest years so, inevitably, he looked to see an opportunity in that direction. He had learned to drive on his parents' property when ten years of age and recalls driving the family Ford V8 down the Great Northern Highway when he could scarcely see over the steering wheel. He also drove an ex-army one-tonne truck about the farm. He acquired his own motor vehicle, a 1937 Hillman Minx, when he was 17 and has since owned a succession of cars ranging from an FJ Holden to an XJS Jaguar. These days he is committed to his Porsche 928S4 which he says is the one car he never gets bored driving.

'You don't have to speed to have fun with a Porsche', he says.

Although Ralph Sarich wanted to be part of the automotive scene, his approach was anything but conventional. He had long harboured a deep conviction that he could improve on conventional technology. He was also keenly aware that the auto business was growing rapidly and offered an unprecedented opportunity for new and potentially lucrative ideas.

The sheer fantasy of the situation never dawned on him. Here was a young, relatively untrained man with little practical experience in automotive engineering planning to take on the mighty engineers in Detroit, Turin, Birmingham, Stuttgart and Toyko, and make his fortune by improving on their designs.

Unconcerned by the magnitude of the task, Ralph decided that conventional car engines and transmissions were too large, too heavy and too expensive and that he would do something about it. To pursue this dream, he needed to make money, so he continued to invest in real estate and bought a few small businesses including a couple of retail shops.

Pat meanwhile kept the home fires burning, working around the orchard and taking jobs with local firms when available. She has a natural aptitude for mathematics and can add, subtract and multiply in her head like a human computer.

'Pat can add columns of figures faster than anyone I have seen in my life', says Ralph.

'Quite phenomenal speed — and accurate too. She can go down five long columns at lightning speed and put the answer down right every time.'

Pat laughed when this quote was read to her.

'When we had the BP station, Ralph used to bet the auditors that I could add up their columns of figures faster than they could', she grinned.

'He never lost.'

For many years, they both worked very long hours. Often Ralph had no more than three hours of sleep at night and Pat raised their children — Peter and Jennifer — and worked extremely hard around the orchard where they lived.

'I was a little girl who had worked in an office all my life and suddenly I found myself picking grapes or lumping rockmelons in temperatures above 100 degrees [38 degrees Celsius]. To help with the family finances, I also took a job operating a ledger machine with the Swan Valley Cooperative. Later, when we took the BP service station, I handled the books and sold the petrol while Ralph worked in the engineering shop.'

Pat says that Ralph often worked late into the night, always staying until the job was finished. Despite the heavy workload, he maintained his love of sport, especially karate which fascinates him. These days he plays tennis on his own court, jogs around a nine-kilometre course near his home and takes frequent workouts to stay fit. During 1988 he completed a course in scuba diving.

When first married, Ralph and Pat lived in the former Sarich family home at Herne Hill, then rented a small home close to the BP service station at Midland and later built a four-bedroom brick home at nearby Cavisham. When Ralph decided to quit his job and devote himself exclusively to automotive research, Pat agreed that they should sell their house (it was just one year old) and use the money to fund research into a lightweight automotive engine.

'We felt that if I failed she could always go back to work as a bookkeeper and we could do what everyone else does: put a deposit on a home and pay it off again', he later said.

'Fortunately, things didn't work out that way.'

They rented a fibro home in Morley, on the outskirts of Perth, which was nowhere near as modern and luxurious as the home they had sold. But it had a large workshop and plenty of land around it and would be an ideal base for the new Sarich enterprise. The new firm was a sink-or-swim venture, especially as Ralph was still unsure exactly what type of engine he would build. The rather vague plan was to design a small rotary unit, prove it could work and sell the idea to a major car-maker. Countless thousands of other inventors had shared the same idea — but few have had the same tenacity, intuitive flair and family support.

'Without Pat, none of this would have been possible,' Ralph now says. 'She was prepared to commit her house and security to my dream. Although not at all interested in engineering, she just wanted me to have the chance to fulfil my ambition.

'She's like that. The only thing she won't let me do is learn to fly. She reckons I might kill myself. I once booked some lessons but she went and cancelled them. I tried it again, and she cancelled them too. It has become a kind of game with us.

'But, apart from flying, when I've ever wanted to do something, Pat has agreed and backed me all the way.'

These days the Sarichs are a close-knit family, with Peter and Jenny, now in their early 20s, working in the family private company which has extensive real estate interests built up over the years.

Sarich gives full credit to Pat for raising two likeable, well mannered children when he was often at work before they rose in the morning and by the time he came home they were in bed.

'Pat used to insist that I take a Sunday off now and again, so we could go for a family picnic', he says.

The four have remained very close, but the days of making every dollar count are over. Pat recalls the day in the mid-1980s when Ralph came home from work, extremely pleased with an agreement he had signed.

'You can each have any car you want', he told the family. Jenny chose a Ferrari, Peter a Porsche. Pat received a Mercedes sports car to replace her Triumph Stag convertible.

According to Pat, Peter has his father's driving ambition but is business oriented and not very mechanically minded. Jenny did a science degree at University and is very strong technically.

'She's also the best lady driver I have ever seen', says the proud father.

'When we went to a Porsche day at Wanneroo Race Circuit, she beat all the Porsches in her Ferrari and held the lap record for the day.'

Both Jenny and Peter are now married and Pat and Ralph live in a three-storey home on the edge of the ocean. The house and cars are the only signs that the Sarich family ranks amongst Australia's richest. Independent reports put the family wealth between $350 and $400 million — a figure which will multiply rapidly if the engine makes the expected inroads into the world's auto business.

Pat and Ralph also have a secret hideaway, land south of Perth on the ocean, where wildlife thrives in a natural sanctuary. The land had been bought by the family company with a view to developing it commercially but Ralph and Pat liked it so much they acquired it. Ralph calls it a 'property' but Pat says it is a well-earned indulgence. He has worked extraordinarily long hours all his life, never looking at the clock, seldom recognising the weekend. This is the one place he can switch off totally and relax, she says.

'I went into the engine project believing I could get quick results', says Ralph when asked about the long hours.

'I had little idea just how conservative the motor industry is. The more progress I seemed to make, the more I became aware that I would have to produce something sensational to even make an impression on the major producers.

'The engine project has taken much longer and needed a great deal more work than I had ever thought possible, but I had to keep going. This meant working 14 hours a day, seven days a week at times.'

Thanks to his apprenticeship days, he became a skilled machine operator. Even today he can use virtually any of the machine tools in the Orbital Engine Company's high-tech facility at Balcatta and takes pride in being a 'hands-on' engineer as well as the managing director. Despite his lack of formal training, Ralph has an intuitive knowledge of sound engineering principles along with a complete refusal to accept established ideas as gospel.

He is, however, willing to admit that he provides the creative input and that it needs professional men such as his chief engineer, Kim Schlunke, and his team, to turn his ideas into reality.

The enterprise (which was employing 168 people by January 1989) started in March 1970, in a small workshop alongside Sarich's rented home. At first, Sarich had a head buzzing with ideas but little idea where he intended to start. He sat in front of a drawing board with a blank sheet of paper, somewhat overawed by the magnitude of the decision he had taken. His security, and that of his family, depended on his ability to design and sell a concept which was still in his head.

His first thought was to develop his oil-based automatic transmission which worked on the hydrostatic principle but took an original line. With the usual hydrostatic system, oil is pumped to small hydraulic motors which drive the wheels, and the speed with which the oil turns the motors is varied by changing the stroke of the oil pump. This means that the oil velocity (and noise level) goes up as the road speed increases. Sarich took the opposite tack and devised a system in which the oil flow is severed when a one-to-one flow is achieved. Effectively, it works in reverse to most hydrostatics and the oil flow decreases with road speed. The noise is quite high at slow speeds and falls as the vehicle gathers pace and, in theory, at a one-to-one ratio it is noiseless.

'It does not require a torque convertor nor even gears and was first fitted to my Falcon back in 1968', he says.

'It worked but you could not call it successful in terms of commercial viability', he says.

After setting up his new business, Sarich considered developing that transmission into a production reality and took the idea to Ford Australia. Engineers here showed little interest, but in the course of the conversation, they mentioned the growing interest in lightweight rotary engines. They expressed a possible interest in finding a rotary engine which compared favourably with the Wankel which was then still under development and considered promising by some engineers. Sarich took up the challenge.

'I had long believed that car engines are too big and heavy and not in keeping with the modern world', he said in 1988. 'Their words reinforced my determination to build a very compact unit with good fuel economy and, hopefully, lower cost and better exhaust emission levels.

'It was not long before fuel economy and emissions became critical factors and car companies started to pour far more money into engine research than they had ever done before.

'According to one report I saw, Ford spent $200 million to improve the fuel economy of its engines and succeeded in producing only one-tenth of a mile per gallon extra. Others made rapid progress and we found ourselves facing a fast-moving target. What was worse, we could spend only a small fraction of the money that the big firms had to play with.

'Having decided to build a lightweight engine, I studied all sorts of different geometries for a rotary — and finally came up with the Orbital concept which I considered to have unique gas motion characteristics. Experts did not agree with me at the time and said the design could not produce good combustion characteristics but they did not know about some tests I had already done on a two-fluid fuel-injection system.'

The new engine was called Orbital because the piston does not rotate about its own axis (as does a rotary engine) but follows a line best described as a closed path.

Despite frequently aired criticisms that the Orbital engine seems to have been around for years, the exercise was done very quickly by industry standards.

Having decided in 1969 to produce a rotary-type engine, Sarich spent three months producing working drawings and ran into considerable difficulty devising a way to actuate and seal the vanes. After one year in business, he was working 18 hours a day, seven days a week building the engine. The entire unit was made on a lathe and milling machine in his workshop as there was no money for specialised sub-contractors.

'In the early days I did the designing, draughting and machining myself. Then I employed Bruce Fairclough, Ken Johnsen and Colin Pumphreys. As the workload grew, the small firm took on its first apprentice, Ralph's nephew, Neil Sarich. The growing wages bill soon put a strain on the slender finances.

'We put out signals that we needed help. Sir Charles Court [Premier of Western Australia] heard about the operation and sent some government officers to have a look.

'One was Tony Constantine, a professional engineer in the steel industry. When he saw what I was trying to achieve, he insisted on making a personal investment and acquired a 10 per cent holding. A friend of mine named Roy Young (who ran an earth-moving company at Carnarvon) also wanted to buy ten per cent and between us we funded the whole project. The engine fired for the first time on 18 June 1972, just under two years from the time the concept was born. It had cost us about $60,000.

'From then onwards, it was a partnership. These days Tony Constantine is chairman of Sarich Technologies Ltd and Roy Young is the second-largest shareholder outside the Sarich family.

'On 23 November 1972, we signed an agreement with BHP. We had several approaches from other people but decided to go with BHP because Tony Constantine had a good relationship with some BHP executives through Wundowie and he thought that BHP had some very good people. As things turned out, they've never let us down and have proved excellent partners.'

The Orbital engine never went into production. Just at the time when General Motors wanted one installed in a car for serious testing, Sarich and BHP decided to concentrate the entire research effort on the new injection and combustion process.

Kim Schlunke, now Sarich's chief engineer, believes the Orbital engine could still be developed into a viable proposition, but as of late 1988, Sarich has no intention of returning to it.

'I knew we could demonstrate the engine to the satisfaction of car companies,' Sarich told this writer, 'but it would take too long for them to develop the tools to mass-produce it. The lead time was too long for our investors.'

These days Sarich operates from a suite of offices in a high-rise office block in St George's Terrace in the heart of Perth's financial district. The office is 20 minutes by car from Balcatta where the research plant and 168 employees are located in a building owned by Ralph Sarich and Tony Constantine.

Sarich is likely to arrive at a business meeting in his silver Porsche, smartly dressed in a conservative suit and tie. His slightly greying hair is well groomed. He smiles easily and often and is relaxed in the presence of strangers as well as with his own people. The boyish shyness and naiveness which were features of his television appearances and press conferences during the 1970s have gone. They are replaced by the assurance of a man who knows he has succeeded on his own terms and can set the rules for any game the media or big business wants to play.

Sarich has an enormous capacity to remain unhassled regardless of the pressure around him or the long hours being worked.

'People often tell me I don't show signs of stress', he said when asked if he had any special techniques for handling pressure.

'I find jogging very relaxing. I often get up early in the morning for a long run in the sandhills. I usually go about nine kilometres but I'm a lot slower now than I used to be.

'If I have a problem, I think about it when jogging. That way, the jogging ceases to be monotonous. Somehow problems never seem so bad when you're jogging and it helps me relax. But I never really get worried about things which might happen. Often I actually look forward to a problem because I like to be mentally active.'

Quietly spoken, cool-headed and thoughtful, he approaches a business or engineering problem analytically and methodically. He was able to persuade BHP and later several major US companies to take him seriously because he put an enormous amount of work into preparing a detailed presentation before making an approach.

His mind works with extraordinary speed. BHP's Dr Robert Ward once summed Sarich up by saying:

'He is like a computer – you can almost see him thinking. When you have a problem you are usually satisfied if you can get one solution. Sarich comes up with 20 or 30.'

He also goes into a discussion or negotiation with the answers to almost any kind of question likely to be thrown at him. Furthermore, he backs his claims with documentation which cannot fail to impress his audience, be they company managers or professional technicians.

By late 1987, Sarich had withdrawn from the day-to-day running of the firm to concentrate on licence negotiations. He is chief negotiator and handles the financial decisions but is not formally qualified in either area. He says it has been a matter of self-education, good fortune and experience.

He says he is not driven by money.

'I try to do the best for people who have shown faith in me', he told this writer. 'I also get a lot of self-satisfaction trying out my ideas. It's a matter of wanting to achieve something in life.

'I want to achieve something extraordinary.'

Sarich has received several awards since he made headlines. He was named Citizen of the Year in 1972 and selected as the 1973 Jaycees 'Outstanding Young Australian'. Further awards included the Sir Laurence Hartnett Award for Technology in 1972, the BRW Business Award for Technology and the Medal of the Royal Society of Arts (London) in 1986. He received an honorary degree as Doctor of Science from Murdoch University in 1987. In 1988 he was appointed an Officer in the General Division of the Order of Australia.

In October 1988, Ralph Sarich received one of the most prestigious engineering awards available anywhere. He was given the 1988 Churchill Medal by the Society of Engineers of London for 'his work over more than 16 years in conceiving, designing and bringing into production his advanced two-cycle Orbital Combustion Process'. This was the first time anyone outside England has won the award.

In making the nomination, the adjudicating panel commented: 'Starting as an apprentice fitter-and-turner, Mr Sarich through a combination of perseverance and ingenuity that is the hallmark of the true engineer, has made an outstanding contribution to the engineering profession worthy to follow in the footsteps of the first Churchill Award Medal winner, Sir Frank Whittle'.

Other recipients of the medal are Sir John Cockcroft (atomic energy), Lord Hives (of Rolls-Royce) for technical education, Sir Geoffrey de Haviland (aircraft), Sir Christopher Cockerell (Hovercraft) and Professor Gambling (optical fibre technology).

The award is the most senior that the Society of Engineers bestows. Approval to use the name Churchill was given by Sir Winston in November 1946. The medal is 35 mm in diameter and carries the name and date of incorporation of the Society of Engineers encircling the name and date of founding of the Civil and Mechanical Engineers Society.

The recipient's name is engraved on the reverse side.

8

The business side

The Orbital Engine Company had its beginnings in 1969 with the establishment of Sarich Design and Development, operating as a one-man business in a small shed in the Sarich family's rented home at Morley, an outer Perth suburb.

As soon as he had the design of the Orbital engine on paper, Sarich rented a small factory in nearby Rudlock Road to have room to install extra machinery. He planned to machine the entire engine in-house. Once it was assembled and running, and BHP had taken a financial interest in the project, the Orbital Engine Company Pty Ltd (OEC) was formed. BHP held half the shares and Sarich interests the other half. In the same month, January 1973, OEC acquired Sarich Design and Development which continued to do contract work to earn extra cash while OEC concentrated on the engine. The BHP involvement meant the company could afford more spacious premises, and within months, OEC moved to a much larger factory on an industrial complex just down the road.

Ralph Sarich meanwhile had formed a partnership between himself, his wife, Pat, engineer Tony Constantine and businessman Henry Roy Young. Tony Constantine (now chairman of Sarich Technologies) is a professional engineer and former general manager of Wundowie Iron Works. He had been given the job of restructuring the Chamberlain tractor works during the late 1950s by the State government and was asked to study the Orbital engine at a time when Sarich was seeking financial help. Constantine was so impressed that he offered some of his own money to help with the research. Roy Young was a long-time friend whom Sarich had met in Carnarvon in northern Western Australia when working for Tutt Bryant.

In 1973, the Sarich interests, BHP and OEC entered into an agreement under which the Sarich partners granted OEC an exclusive licence in relation to the Orbital engine in return for royalty payments. BHP agreed to fund the development program and, in return, became entitled to the profits of OEC after deducting the Sarich partners' royalty payments.

Three years later, effective from June 1976, this agreement was amended and the deed recorded that all research into and development of the Orbital engine would be carried out by BHP at its own expense in accordance with programs established by OEC. In other words, BHP was to fund the entire operation.

About the same time, a trust — in which private investors were invited to purchase units — was established to acquire the interests of the Sarich partners in both the amended agreement and the partnership. The trust was originally called the Sarich Design and Development Unit Trust and created by a deed dated 15 December 1978. (Six years later it was renamed Sarich Technologies Trust.)

The unit trust had been set up so that any income receivable would not be taxed at company level but after distribution. This was a tax-effective way of ensuring its earnings would be taxed only after reaching the hands of unit holders, an idea which appealed to those tax-free institutions which had invested in it. After this particular avenue of tax minimisation was closed by the Federal government, the advantage of remaining a trust disappeared and the operation was converted into a conventional limited liability company, Sarich Technologies Ltd, in November 1988.

A public company is more flexible in its ability to retain earnings or raise capital by borrowing through acquisitions or by mergers than a trust.

Meanwhile, in 1979, the firm had moved to Balcatta, another Perth suburb, to a site owned by the State government. Tony Constantine and Ralph Sarich bought this land from the government and built a new research facility for OEC. The same building, in expanded form, is still leased by OEC as its headquarters.

The Orbital Engine Company remains Sarich's main research organisation and is effectively a private firm owned by two public companies — The Broken Hill Proprietary Company Ltd and Sarich Technologies Ltd. The latter has approximately 12,000 shareholders.

OEC has developed into one of the world's leading organisations devoted exclusively to internal combustion engine research. Even if it did not have its own technology to sell, OEC would be a viable commercial proposition able to undertake basic research on behalf of existing engine manufacturers. It has both the facilities and expertise to do well in the field.

Sarich has very successfully built a young and imaginative team of highly qualified engineers who can turn his ideas — which are frequently brilliant — into reality. They do this the old-fashioned way, with long, hard and dedicated work. The entire staff — numbering 168 in early 1989 — arrives early and frequently stays very late. Working on Saturday morning or afternoon seems to be the rule rather than the exception.

The organisation is informal and flexible, with the managers reporting to Ken Johnsen who reports to Sarich. However, everyone in the organisation has direct access to the boss because he maintains an open-door policy. The employees are keen. They walk briskly throughout the facility, and if they talk for any length of time, it is invariably about business. One of the many achievements which show the team spirit was the cooperative design of a vehicle which achieved a world record of 2685 miles per gallon in the Shell Marathon of 1980. This was designed and built privately during the employees' own time and it was their decision to put the company logo on the car.

The 9000 square metre OEC facility is equipped with approximately $20 million worth of scientific, testing and other equipment, including three CAD/CAM work stations, computerised cuttings machines, emission laboratory, laser particle measuring device, seven engine dynamometers, two marine engine dynamometers, two chassis dynamometers and an extensive machine shop. There's a large electronics section where engine management systems are designed and built under the direction of Peter Simons, one of the company's most senior engineers.

Simons joined Orbital in 1977 after graduating from the University of Adelaide with degrees in Electronic Engineering (hons) and in Science. Initially he designed OEC's high-speed and low-speed data acquisition equipment and was then responsible for designing and building the engine management control units (ECUs) for use in vehicles. Over the years, OEC's electronic department has developed many marketable products.

OEC's policy is to hire engineers straight from university, while they are still young and enthusiastic. The engineers are kept informed about engineering developments throughout the world by means of an extensive library maintained on database. This was a far-sighted but expensive concept when instituted during the early 1970s but has since proved invaluable.

Until mid-1986, the operation was funded by BHP. Since then, licensing and other agreements have brought in revenue as high as $6 million per client.

Manager Ken Johnsen runs a tight ship which includes a worldwide patent library on database. He joined Sarich Design and Development in March 1972, just before the Orbital Engine Company was set up. Starting as a trainee engineer, he acquired practical experience operating the machinery and theoretical knowledge by attending college part-time. Jobs for would-be engineers were thin on the ground in Perth at the time, but Ken happened to be a neighbour of the Sarich family and his parents knew Ralph and Pat. Ralph was looking for his first engineer but had not met Ken until he arrived for an interview.

Like most employees in a small factory, Ken Johnsen did just about every job around the place, including machining, assembling and draughting. He had a stint in charge of the fitting shop and later as assistant chief design engineer. The chief designer at that time, John Hinton, is now OEC general manager, reporting to Johnsen.

During the early 1980s, Ken moved from the hands-on area and became a commercial engineer, liaising with customers and potential customers. Ralph Sarich and engineer Kim Schlunke operated as the main sales team, taking their technology around the world as they looked for expressions of interest. Ken stayed at home organising the patents, legal agreements and office staff. He was appointed manager of OEC in 1986 and now controls the whole operation, reporting directly to Ralph Sarich. He says half his time is spent managing people and half negotiating with lawyers. The licensing agreements increasingly involve overseas companies. In 1988 alone, he visited the USA five times and made additional trips to Japan and India.

Kim Schlunke, who heads up the mechanical engineering staff, was born in the NSW town of Glen Innes where his father had a farm. After leaving school, Kim joined Tubemakers of Australia (now a BHP subsidiary) as a cadet and studied engineering at the University of Newcastle, NSW. After graduating, he was in the process of doing a Masters degree when he married and travelled with his wife to Perth for a honeymoon.

His visit coincided with one of those rare occasions when a Perth firm advertises for a graduate engineer. The firm happened to be the Orbital Engine Company and Kim was intrigued by the advertisement as he had heard Sarich and his team were engaged in adventuresome research. He sought an interview and was hired in April 1977, working alongside Peter Ewing, a graduate engineer doing fundamental research on the Orbital engine.

Another well-qualified employee was Dr Dominic Swinkels (his PhD was in chemistry) who was director of research from about 1974 to 1979 but, despite its heavy financial commitment, BHP never sought to look over Sarich's shoulder and left all engineering and negotiation matters to him and the OEC team.

Working under Swinkels initially, Kim made rapid progress within the organisation and ultimately became manager of engineering. He currently sits on the board of management and the Orbital-Walbro manufacturing venture.

'The firm provides a fundamentally very satisfying working environment', he says. 'Any reasonable idea which is suggested is quickly tried. We pride ourselves on our ability to 'turn through' experimental devices very rapidly.

'Some experimental organsations are heavily endowed with boffins, but we are heavily endowed with thinkers and creators. We also have a large number of doers and makers, our machinists and technicians. We can "cut and try" a lot faster than most other companies and we tend to see new experimental hardware in our hands a lot faster than is possible elsewhere.

'We also have a very open forum for discussing new ideas with lots of formal and informal meetings.

'On the downside, there's lots of pressure. We are trying to do something that no one else in the world has done before, with the possible exception of the Wankel people, and we are doing it with only 168 people. At times, we have to deal with a complex engineering problem which, in a major car company, would involve an entire department.

'That makes the work challenging and interesting but also creates a load of pressure.'

In July 1988, Kim Schlunke won the Rolls-Royce-Qantas Engineering Award for excellence, receiving the Warren Centre Medal and a $20,000 cash award. This annual award is presented to the person judged to have done the most to exemplify engineering excellence in Australia. He was chosen for the award after the adjudicating panel had spoken to major car companies and marine engine companies. The panel learned from them of Kim's vital role in the conception, development and marketing of the Orbital Combustion Process (OCP).

In 1987, Sarich started to withdraw from the day-to-day running of OEC to concentrate on licence negotiations. He remains, however, very much in control of the entire organisation despite his concentration on the commercial rather than the engineering side of the enterprise.

'As an engineer, he is excellent. No one can touch him for motivation and dedication', says Kim Schlunke. 'But as a businessman, he is second to none.'

The family business

The Sarich organisation is very much a family affair. Pat and Ralph Sarich have always operated their enterprises on a 50/50 basis.

They each originally had a 22 per cent share of Sarich Technologies Trust but Pat sold some of her holdings in 1987. She now has a 19.42 per cent holding which is still three times larger than the next biggest shareholder, ANZ Nominees Ltd.

Although Pat has worked hard and long on the financial side of various family businesses, she is now content to keep abreast of what is happening without taking day-to-day responsibilities.

The Sarich offspring, Peter and Jennifer, both in their 20s, are heavily involved in Sarich Corporation Pty Ltd, the private family company which has numerous investments and is a major property developer in its own right.

Peter is managing director and, according to his mother, is every bit as ambitious as his father was at the same age. Unlike Ralph, Peter is not very technically minded but has a strong grip on business practice and principles. As of late 1988, the private company was building Perth's most luxurious home units at Nedlands on the Swan river. It was also engaged in property development in the suburb of Morley and creating a shopping centre in Cairns, Queensland, located next to the Park Royal hotel. The family company had planned a 200-hectare holiday resort at a fashionable ocean township, adjacent to a national park, but Pat and Ralph fell in love with the location. They transferred the title to their own names and turned it into a private retreat.

Jenny Sarich is the only member of the family with a university education (she has a BSc and, in 1989, was studying for a Master of Business Administration degree). Jenny is married, has a strong interest in technical and medical subjects and carries numerous business responsibilities. Ralph described her as 'the keeper of the family funds' adding that she is meticulously accurate, level-headed and shrewd. She is honorary secretary of the Medical Foundation of WA (Medwa), the Sarich-funded research organisation, and acts as the coordinating and administrating officer. Her duties include sorting out all applications for research funding and arranging for a scientific committee to evaluate those which show promise. She also keeps tabs on all expenses incurred by Medwa.

Interestingly, Ralph's well-known edict about not drinking on the job includes his family. He has an inflexible company rule that no one — including his son and daughter — may drink alcohol during office hours or when on company business. This includes taking a client to lunch and the penalty for breaking the rule is instant dismissal. Ralph leads from the front. Although he enjoys both beer and wine when off the job, he never touches a drop when involved in business of any kind.

'I believe you cannot drink alcohol and operate effectively at the same time', he says.

'It is counterproductive and potentially dangerous. I know from my early days working with a heavy machinery firm that if I went with the bosses to a beer garden for lunch, it used to virtually write off my afternoon.

'I recall an occasion when I felt contempt for a certain boss when he was telling me what to do while smelling strongly of alcohol and slurring his words.

'Besides, within our own research facilities, we have skilled men machining very expensive items of prototype components – one error could injure the operator or damage the component. The risk is not worth taking. But how could I have a company rule saying that men operating machinery could not drink at lunchtime but executives entertaining clients could?

'So I decided there will be no alcohol for me or anyone else on the payroll during working hours. The rule applies to all our companies and organisations and there are no exceptions.'

That he is a strong – if fair – boss is accepted through the company.

'Ralph pushes himself and everyone else pretty hard', says Ken Johnsen. 'He drives by example and this makes him a hard taskmaster.'

Ralph himself never seems stressed or tired, no matter how hard he is working, and tends to think others are as durable. One day when the author was interviewing him in his office, the phone rang. One of his senior executives was at the other end and Ralph listened for a while, then grinned:

'What do you mean, you're overloaded?' he asked. 'I don't know what the word means. You're nearly half my age and I'm looking around for more things to take on! How could you have too much to do? Just take a deep breath and get into it, my lad.'

He looked at me, smiled and shrugged as though to say 'They don't make them like they used to'.

When Ken Johnsen was asked if Ralph was a tough boss, he chose his words carefully, then said:

'When someone gets up to leave at the end of the day, he's likely to look at the clock. Sometimes he finds it hard to understand that some people are just on wages and do not want to work the hours he and many of the senior people put in.

'On the other hand he is generous, both personally and in a business sense. He's well aware that he drives the staff hard and this is why he has set aside about 5 per cent of his holdings which, on current valuations, means $14 million. When he deems the company to be successful, he will distribute that money to the people who have helped to make the company.'

Major OEC agreements

In May 1974, an agreement was signed with the Sydney-based company Victa Limited to develop a small Orbital engine for use in its motor mowers. This was the first agreement of its type negotiated by Ralph Sarich, despite the headlines of the day which suggested he was beating off offers from multimillion dollar companies. After Victa had spent a considerable amount of time and money, the board concluded that the engine would be too expensive for its intended use and the licence was relinquished.

GM was rumoured to have spent nearly one million dollars in total with OEC during the Orbital days and the first really promising automotive agreement Sarich signed was a deal with GM in 1983 to produce a prototype of the Orbital engine. That was OEC's first written contract in which money actually changed hands. GM set a number of targets which were to be achieved before they would sign a licensing agreement. In the end, the deal never went ahead because OEC reassessed its own future following the rapid developments taking place with the OCP systems. In February 1984, Phase One of a new contract with GM involving the OCP system was signed. Two months later Outboard Marine Corporation of Waukegan, Illinois, signed an option involving the development of the OCP system for two-cycle marine engines. This was followed by Phase Two of the GM contract in which OEC undertook to fit an engine to a GM car for testing and evaluation.

In December 1984, Mercury Marine Corporation of Fond du Lac, Wisconsin, took out a licence option.

As part of the GM agreement, a Chevrolet Spectrum powered by an OEC three-cylinder engine was delivered to Detroit in April 1986. The agreement provided for numerous penalties should there be a shortfall in technical performance but GM later paid in full, acknowledging that the engine fulfilled all specifications.

In the following month, an agreement was reached in principle with Walbro for the manufacture of the injection system. One month later, June 1986, Outboard Marine Corporation purchased a licence to manufacture and sell Orbital technology engines for marine and selected industrial applications. In September that year, Ford entered into a development and testing program, funded by Ford, in which OCP engines were fitted in Escorts and shipped to Detroit for evaluation. Under this agreement, OEC was required to meet Ford's specifications relating to driveability, fuel consumption, emissions, acceleration and other technical requirements. The vehicle was delivered in September 1986 and Ford engineers later acknowledged that the car met all agreed specifications.

In June 1987, Mercury commenced a test program on a six-cylinder outboard engine fitted with OCP injection. Depending on operating conditions, they found fuel economy gains of between 30 per cent and 70 per cent compared with the conventional version of the same engine. There were also significant emission advantages.

One British, two European car makers and five Japanese auto, marine and motorcycle firms started discussing business arrangements and terms during 1986.

In May 1987, a leading Japanese motorcycle and car engine-maker took out a licence option and Orbital Walbro Corporation was formed and granted a full licensing agreement. One month later, Mercury signed for a full manufacturing licence.

In June 1988, Ford Motor Company of Dearborn, Michigan, entered into a full licence agreement allowing it to make OCP engines worldwide. This was an historic agreement because it meant that Ford had departed from all previous practice and had paid a very substantial amount of money for engine technology produced by an independent organisation. It is worth remembering that Ford bought 20 per cent of Toyo Kogyo (Mazda) in 1979 and has since had access to Wankel rotary engine technology, Mazda being the world leader in the field. Ford has, however, shown no public interest in using a rotary engine.

As of late 1988, Orbital Engine Company has signed three other significant agreements. It is licencee to Outboard Marine Corporation (makers of Johnson and Evinrude marine engine) and Brunswick Corporation (Mercury Marine) and has a multimillion dollar development agreement with General Motors. The Orbital-Walbro partnership, formed in June 1987, involves a $25 million commitment from BHP and an equal amount from Walbro Corporation.

As of early 1989, negotiations are also in hand with a major motorcycle maker, two manufacturers of small engines and five Japanese car firms. The preliminary design of an aviation engine has also been completed.

GM is expected to sign a full manufacturing agreement during 1989.

9

The financial rollercoaster

When shares in the Sarich enterprise went onto the market in 1984, they sparked off a drama which proved no less gripping than the events behind the engine itself. A small engine research company with an initial book value of $2.1 million had, within a short time, a paper value of almost $930 million in spite of the fact that not a single production engine had ever been sold.

The reason is that Orbital has always been considered a technology company by its shareholders and management and its source of income is from the sale of technology not engines. All agreements entered into by Orbital involve the payment of advance fees. In the case of an engine, $10 per unit is payable in advance on the proposed output, so considerable income is generated before production commences.

The company's story is one where investors have showed remarkable faith in Ralph Sarich despite setbacks, delays, ugly rumours and a government report which suggested that it would take at least a further seven years before production could start. The investors ranged from pensioners to financial institutions and had one thing in common: they clung tenaciously to their holdings in the belief that the company would come good.

For those commentators who have hinted that 'going public' provided Sarich with the opportunity to profiteer, it should be added that major financial institutions did not support this view. Sarich has never sold any of his own shares, even when the valuation made him one of Australia's wealthiest men on paper, though his wife sold 25 per cent of her shares. His major shareholders included — and still include —

Australian Mutual Provident Society, Rural and Industries Bank of Western Australia, MLC Life Ltd, Associated National Insurance Company Ltd, Mercantile Mutual Life Insurance Company Ltd and the Superannuation Board of Western Australia. More than a few major public companies would do cartwheels of joy if they could secure the backing of such a prestigious group of shareholders.

Things were not always that way. For many years there was no direct avenue through which the public could participate in Sarich's dream. Potential investors had to wait 12 years and only then could the considerable strength of public support and emotion be properly gauged.

Ralph and his partners launched the Sarich Design and Development Unit Trust in 1978 and, six years later, listed it on the Stock Exchange renamed Sarich Technologies Trust (STT). In doing so, they unleased a sharemarket darling volatile enough for the Poseidon boom which was raging across Australia during the Orbital engine's early gestation period.

The directors of ST Management (STM), which handles STT's affairs (trusts do not have their own directors), have been kept constantly on their toes since the float coping with a potent mixture of wild speculation, misleading rumours, blue-sky optimism and the tantalising possibility of mass production which seemed to ebb and flow depending on who was talking or writing about it.

The effect was to make the STT's unit price graph look like the trajectory of a balloon released before the knot is tied in its neck. STT has become perhaps the most queried company on the Australian sharemarket as the Perth Stock Exchange has regularly challenged the directors of STM about rapid fluctuations in share prices in a bid to minimise the effect of rumours and keep investors fully informed of the trust's prospects. They have skillfully done this without prejudicing the high-level negotiations OEC has been holding with some of the world's most powerful industrial companies.

The stock gyrations have occasionally made Sarich one of Australia's richest men, when measured on paper, though he remained singularly unimpressed by the fact.

Apart from the obvious reason that money was needed to fund the trust's share of future research and development work, there appeared to be two important motivations behind the decision to float the Sarich partners' half of the Orbital Engine Company through the listing of Sarich Technologies Trust. One was to give the public at large a chance to participate in the anticipated success following the support Sarich had received over the years, the other was to create a means whereby his employees could acquire shares which would give them a stake in their own future and a share of the wealth their labours created.

The original trust, the Sarich Design and Development Unit Trust formed in 1978, could have been floated much earlier than 1984. There had always been public pressure on Sarich to 'go public' but he was reluctant to do so until he was reasonably sure he could offer investors some solid prospects. Previously, the only way the public could invest in OEC was to buy BHP shares but this was impractical as the OEC investment was a very small part of the total BHP picture. For every $100 invested in BHP in 1978, only about 50 cents represented a piece of Sarich.

Before the float, STT's sizeable stake in OEC was worth, in strict accounting terms, a modest $2.1 million as OEC's balance sheet showed net assets of $4.2 million at May 1984. This reflected the fact that, from 1976, BHP had been solely responsible for funding the development of the Orbital engine and, therefore, these outlays did not appear on OEC's balance sheet or in its profit and loss account. However, as OEC held the patents, the improvements made to the engine and the other technologies increased the value of the patents and, therefore, of the company. This was reflected in the transfers of units in the unlisted trust between 1978 (when it was formed by the Sarich partners) and 1984 when it was floated to the public.

After an initial valuation of $6 million in December 1978, the trust's value actually went down a little. Sales were made person-to-person, not through a stockbroker, because the trust was not listed on the Stock Exchange at the time. Some early sales of units were made at prices which inferred a value of just $4.5 million on the trust. Few, if any, units were sold during the next two years but, in March 1981, some changed hands at prices which inferred a value of $10 million for the trust. Clearly, development was making headway and the prospects were looking better.

As research progressed at OEC, so the value of the trust's stake in the research organisation rose. In the early part of 1982, unit prices in this off-market trade indicated the trust was worth $15 million and, in June that year, sales inferred a value of $20 million. This figure had risen to $26.8 million by July 1983.

This rise in value was not officially recognised in the trust's own accounts as the assets were still at their 1978 values. So before the float was made, the directors of the trust's management company, ST Management, officially revalued the stake in OEC and its patents in order to derive an accurate value for the trust's main asset. An increase in value of $23.9 million was decided upon and added to the trust's balance sheet, thus lifting the total value of the trust to $30 million.

The trust had only 6000 units on issue, each issued at a price of $1000 in 1978. The revaluation of assets meant that these shares were now worth $5000 each. In order to protect their value, any new units would have to be issued at $5000 each, but this was clearly too much to ask when going to the general public for the first time. So the directors divided each $1000 unit into 1000 $1 units, changing the issued capital from 6000 $1000 units to 6 million $1 units.

However, these $1 units were still worth $5 each after the asset revaluation, so the directors adopted a common ploy used when companies or trusts are floated. They made a four-for-one bonus issue to the original unit holders of 24 million units, lifting the issued capital from 6 million $1 units, with an asset backing of $5 each, to 30 million $1 units with an asset backing of $1 each. The value of the trust had not changed, but with the asset backing reduced to $1, the new units could be offered to the public at an acceptable price, making them accessible to the hundreds of small investors the directors hoped to attract.

The directors decided to raise cash by issuing 5 million new $1 units at par value. This parcel of units would lift the trust's total capital to 35 million units and represent 14.3 per cent of the expanded issued capital.

Events proved that the decision to issue the units at par was ultraconservative. The demand was so high that investors would most likely have paid twice as much.

Of course, the issue watered down the stakes held by the 60-plus unit holders immediately before the float. After the float, the proportion held by Ralph Sarich and his associates represented 56.6 per cent of the total units, while the 60 other unit holders shared 29.1 per cent.

The two sharebrokers underwriting the issue, Perth's Hartley Poynton and Company and Melbourne's Potter Partners, had no trouble finding subscribers for the float among their own clients. The issue opened on 21 August 1984 and closed fully subscribed on the same day. This was not solely an expression of faith in Ralph Sarich and his colleagues. The Australian sharemarket at that time was in the grip of 'high-tech' fever and any company claiming to have any sort of advanced product was invariably rushed by investors. The issue did not come close to satisfying the public hunger for STT units and the pent-up demand made its presence felt when the units were listed on the national Stock Exchange.

The price of the trust's units rose steadily during the following months as investors kept chasing the hard-to-get scrip. However, early in December 1984, the market for STT units went berserk as investors or, perhaps, one investor tried to get scrip at any price. The price rocketed from $2.85 a unit on 3 December to $3.85 on 5 December as speculation spread about a promotional tour Ralph Sarich had made to the USA.

The price of $3.85 was almost four times the figure investors had paid for the units less than four months previously. Put another way, it meant investors thought that the Sarich Technologies Trust, valued by experts at $30 million four months earlier, was now worth $116 million even though no announcements had been made since the prospectus had been released.

The committee of the Perth Stock Exchange is charged with the responsibility of keeping the market informed so that investors are not misled by rumour or speculation when they buy units or shares. This committee reacted swiftly and, on 5 December, ordered that a query tag be put on the listing board next to STT's name. This was a warning to investors that the committee was trying to get to the bottom of the price surge, a fairly common precaution, especially in Perth which has many mining companies on its boards. Mining companies are prone to rapid price fluctuations because it is not unusual for rumours to circulate about a company's exploration results before they have been announced to the exchange.

The exchange was soon to learn that 'high-tech' companies were as prone to speculation and rumour as mining companies. Perhaps because they understood relatively little about the technology involved and the automotive industry in general, investors seemed prepared to believe almost anything they heard about STT. In particular, they thought events would move far more quickly than they did. Another important factor in the rocketing price of shares was the simple law of supply and demand. Very few units were ever traded on the market because those who had them were either settling in for the long run or waiting for an even greater price rise.

On 5 December the Perth Stock Exchange served on STM directors the first of many queries related to STT price movements. It comprised three questions:

1. Is the board of the manager (STM) in possession of information which, if generally available to the public, might reasonably be regarded as an explanation of the market price increase?

2. Does the board have any reason to believe or suspect that any person has bought or sold units on the basis of any such information?

3. Are there any matters of importance concerning the operations of the trust that the board or manager is about to announce to the exchange and can any such announcement be made immediately?

The directors of STM were given until 7.00 a.m. the next day to answer. They replied negatively in each case. The statement that no new information was imminent calmed the market and protected the trust's commercial interest, but it wasn't correct.

Just 14 days later, the chairman of STM, Tony Constantine, wrote to the exchange giving some details of agreements signed between OEC and the world's two largest outboard engine manufacturers, Brunswick Corporation and Outboard Marine Corporation. Brunswick, which makes Mercury outboards, had just signed a development and option agreement relating to licensing rights for the Orbital Combustion Process (OCP) engine. In doing so, it caught up with its arch rival, Outboard Marine, which makes Johnson and Evinrude engines and which had very recently secured its own agreement.

Mr Constantine went on to say that General Motors (GM) had decided to move faster than planned towards a more advanced technical and business arrangement involving the installation of OCP engines in vehicles. The OEC and GM managements were also discussing transferring some OEC personnel to Detroit.

In what looked almost like an afterthought on the bottom of the page, the chairman added that, apart from Brunswick, Outboard Marine and GM, seven other engine manufacturers had initiated talks and expressed serious intent about entering business agreements with STT during Mr Sarich's recent US tour.

'Several Japanese manufacturers have also expressed serious interest', he added.

Obviously, these statements were downplayed because they referred to talks of a tentative nature but investors were in no mood to be played down. They applied more buying pressure on STT units and the price continued to rise steadily into the New Year. By early February 1985, the Perth Stock Exchange was again sufficiently concerned to challenge directors with the 'three-question query'.

The STT directors responded on 6 February and this time went a little further than issuing a straight 'no, no, no', although they gave no details. They merely referred to 'commercial discussions with a number of major international corporations' and the 'highly impressive technical progress' made with some of the technologies being developed. They said the trust's first half-yearly report to unit holders would be made later in the month.

Unit holders were heartened when they saw the report issued on 26 February. It said that the OCP engine had exceeded all the criteria laid down for any new engine by 'auto manufacturers'. It was lighter, had more torque, was more economical, cleaner and would be up to $200 cheaper to make than a conventional engine. Not only that, the report claimed that car manufacturers would be able to save a further $200 to $300 a vehicle, making total savings up to $500 a vehicle, thus further enhancing the OCP engine's attraction.

With stimulation like this, the share price kept rising as investors clamoured for scrip which was, effectively, unavailable because no one was selling. By the end of March 1985, the few units that changed hands were trading for an amazing $6.80 each, valuing the trust at $238 million and the Sarich associates' stake at a cool $135 million. On 2 April, Ralph Sarich announced he had decided to extend the ownership of the trust to include employees at the Orbital Engine Company, unveiling a scheme under which 1.4 million units would be made available from the Sarich family's own holdings.

The buying pressure on the units continued, and little more than a week later, on 5 April, the price closed at $8, up 50 cents for the day after briefly touching a heady $8.90. The $8 closing price valued the trust at $280 million, comfortably elevating Ralph Sarich to the 'rich rich' club. This enviable situation was dismissed by him (in an August 1985 interview) when he took exception to society trying to judge people's status according to monetary values. He said he would rather be recognised for having beaten the world's biggest car makers at their own game. He added that it was more exciting to know that his OCP engine was being tested in Detroit.

On 16 April 1985, in an address to the Securities Institute of Western Australia, Sarich for the first time indicated the size of the royalties that OEC was negotiating with major car makers. He said he was looking for something between US$20 and $30 an engine which seemed a modest enough target when the technology promised an overall saving of US$500 per car. Meanwhile, the share value had dipped a little after hitting $8 in April, but the disclosure of the size of the potential royalties, innocent enough in itself, was sufficient to push up the units 30 cents to $7.60 each on 18 April.

No great news came from the trust during the next six months and the unit price had eased to around $6 by the middle of October. Two months earlier, the trust manager had released the year's operating results, but as the only income was the interest on the $5 million raised in the float, there was no excitement to report. The trust had made a profit of $270,000 for the 12 months to June, representing less than one cent for each of the 35 million units on issue. Not surprisingly, the result was ignored by the market.

The annual meeting in October 1985 served once again to focus attention on the trust and the unit price started to regain some lost ground. By 6 November, it was back to $8 and, after dipping a few cents, started to surge again, racing past $9 in early December and touching $10 a unit before closing for the year at an even $9.

This meant that the trust was valued at $315 million for the start of 1986. It should be remembered that the general sharemarket was then in the middle of a five-year boom (which ended in October 1987 following the worldwide sharemarket crash). But nowhere else on the boards had the STT performance been equalled. Speculators noted that the trust's units had rocketed from $1 when floated in the second half of 1984 to $9 just 15 months later — an annual growth rate of 640 per cent.

The pressure eased a little in the quiet January 1986 trading period but investors returned in February, pushing the price back to $9 and then towards $10.

The Perth Stock Exchange was alarmed again during April when the price jumped a full dollar to $10.50. Instead of issuing another formal query, the exchange suspended the units from trading and released an odd statement saying they had been suspended for the afternoon session 'pending release of a statement which we have been informed will not reveal any fundamental developments in the trust's activities'.

STM directors replied they believed the surge in price had been sparked by the simultaneous recommendations of STT units by several sharebrokers. Their recommendations must have been persuasive because a price explosion followed. Once the units were past the gravitational pull of $10, it seemed that the stratosphere was the limit. The $11 mark came and went in April, as did the $12 and $13 marks during a matter of a few weeks.

If the STT unit price was already heading into space, an announcement on 9 May served as a booster rocker to take it to a new orbit. BHP — the other half of the OEC equation — released details of OEC's first manufacturing plans. It revealed that OEC was to enter an agreement with Walbro Corporation of Cass City, Michigan, to start a 50/50 joint venture to manufacture injection systems, engine management systems and related components for use with automotive and marine engines in North America and Europe. It was appropriate for BHP to make the announcement because the company was required, under its earlier commitment to OEC, to fund the first $50 million of any investment in a manufacturing plant.

The Walbro deal was seen by many as the move which opened the floodgate, as many potential licencees for OEC's technology had been pressing for a reliable supplier to make the fuel systems and engine management components they would need for any OCP engines the licencee made. The news appeared to get out early, or someone got lucky because, on 7 May, two days before BHP broke cover, STT units were selling for an unbelievable $16.80 each. After a brief respite, the figure took off again, reaching $17.50 during the third week of the month.

On 10 June, STM gave unit holders more good news. Outboard Marine – which already had a licence option – had taken the plunge and signed a licensing agreement for OCP's technology, allowing worldwide manufacture. STM directors were coy about the details but said 'front-end payments and royalty fees are fair and appropriate to the magnitude of the technologies involved. OEC should derive considerable benefits from this relationship.'

The directors also pointed out that the Walbro joint venture agreement, announced a month earlier, was ready to proceed. In addition, negotiations with other manufacturers in the vehicle, industrial and marine areas had reached advanced positions, even the concluding stages in some cases. They disclosed that the proposed licence fees now ranged between US$25 and $60 per engine while front-end fees would be measured in millions of dollars.

Furthermore, the directors added, OEC had been invited to form a joint venture in the USA, and separately in Europe, to make OCP engines for vehicle, marine and industrial uses. In an almost throwaway line, they added that the engine could be adapted to diesel specifications.

The STT units raced to $20 each during mid-June as the news sparked another round of feverish buying, valuing the trust at an amazing $700 million. In round terms, this meant that it was worth 20 times as much as in late 1984 and that the Sarich family now held investments worth about $350 million. This news was not, however, a source of pleasure to Ralph Sarich who continued to be concerned that the small investors, those who had shared his dream from the early days, were being frozen out by the high price just as the dream looked like coming true.

The problem became worse when more important news hit the streets on 4 September. The STT directors announced that the world's number two car maker, Ford, had decided it had better check out what GM had been studying for some time and had entered a program whereby an OCP engine would be installed in a Ford car and shipped to Detroit. Furthermore, this would not be a long drawn-out affair as the test program was scheduled to last for ten weeks after the car arrived and the parties intended to negotiate a manufacturing licence within two months provided that the testing proved satisfactory.

Having eased back from their June peak, the STT units raced to $19.50 each. When it was reported that Ford had taken out a licence option (as opposed to a full licence), the units surged to $21.50.

Ralph Sarich and his fellow STT directors decided to do something about making the units more accessible to the man-in-the-street. They came up with two ideas. The first involved the STT shares held by Mrs Pat Sarich and her interest in things medical. The Sarich family decided to sell down her direct interest of 26.14 per cent to 19.71 per cent by releasing 2.25 million units onto the market in a bid to satisfy demand and make the stock more liquid and available.

These units were sold at $19.95 each, raising $44.9 million which was used, among other things, to make donations towards medical research. Some of the money was retained by the family and some used to meet further capital commitments to OEC, in particular to help pay for an expansion of OEC's laboratories which had always been provided by the Sarich partners, not owned by OEC itself.

When announcing the sale of the units, Mrs Sarich said that, apart from the medical and expansion funding plans, the sale was made in response to requests from overseas institutions and stockbrokers for more stock to be available. The overseas interests had simply been unable to buy stock because trade was so thin. She said this was borne out by the fact that 90 per cent of the 2.25 million units was sold within 48 hours of reaching the market.

The sell-off had little effect on the price which stayed around $20 for the next three weeks. Then, on 9 October, STM revealed the second part of its strategy to make the stock more liquid and the price more accessible.

The directors announced plans to split each $1 unit into five 20 cent units, thus increasing the total number on issue from 35 million $1 units to 175 million 20 cent units. Oddly enough, this plan only heightened pressure, and between the date of the split announcement and the day when the books closed, the price soared to a massive $24, valuing the trust at $840 million.

The split took effect from 31 October, four days after STM had announced that presentations to European car makers, including some in the prestige area, had been very successful. Seven of the nine companies visited had produced letters of intent to enter a business arrangement with OEC while the others expressed their serious interest orally.

It was also announced that Brunswick Corporation (Mercury Marine) had converted its licence option and that the joint venture production agreement with Walbro was being finalised. Directors again indicated that royalty payments were to be higher than earlier indicated and now ranged from A$25 to A$84 an engine, depending on size and application.

All this good news temporarily countered the effect of the five-for-one split. After the new units were listed on the Stock Exchange, the price rose further to peak at $5.30 per unit, the pre-split equivalent of $26.50, and valued the trust at an incredible $928 million. Even recognising that OEC had plenty of positive contracts and licence agreements, it was still remarkable that the trust was being valued at almost $1000 million without a single engine being mass-produced.

The dizzy escalation came to a sudden halt in late November when the units dived from $4.20 to $3.60 each, stripping a tidy $105 million from the trust's paper value. Even though long since lulled by the irrational and erratic price changes, the Perth Stock Exchange was quick to spot the reversal of market sentiment and shot out another formal query.

This time, the STM directors replied:

'The only information we have in our possession which can possibly explain the unit price decrease is that shockingly misleading rumours have been initiated and vigorously promoted, suggesting that a very bad report will be issued by STT in respect of Orbital technologies.

'In view of the directors' intimate knowledge of the Orbital Engine Company's technologies, we strongly deny such rumours and suggest that the only explanation that can be found for these grossly misleading rumours is that it is the result of unscrupulous business activities.'

The directors went on to detail a list of developments including positive viability studies by two governments, a similarly positive study by a major car maker interested in a joint venture engine plant and the start of construction of the Orbital-Walbro fuel system plant in the USA.

In addition, another international manufacturer had started an urgent program to introduce OCP technologies while a further two complete licence agreements, with significant front-end payments, were expected to be completed by January 1987. This release stabilised the market, and by year's end, the units were selling for $4.10 each, equivalent to a pre-split figure of $20.50.

The units continued to improve during early 1987, reaching a high of $5 during feverish buying activity late in January. The price had eased back to less than $4 by April, barely reacting to the STT interim profit which showed a 270 per cent rise to $440,000 for the year. This was up due to a $305,000 lump sum received from OEC, presumably as part of the front-end fees that were starting to roll in.

In May 1987, STM announced that a manufacturing licence option had been signed by a leading (unspecified) Japanese car maker. Two weeks later, on 29 May, STM also revealed that the agreement creating the Walbro joint venture company, the Orbital Walbro Corporation, had been finalised. This came almost six months after the partners had started work on the factory in Michigan and bore out the faith they both had in the potential demand for OCP technology from the marine industry.

The lack of reaction on the sharemarket to these announcements was remarkable, considering the importance of becoming established in the US marine field. This indifference was highlighted by the fact that Australia was in the grip of a rampant bull sharemarket which, unknown to anyone at the time, had only a few months more to run. Prices everywhere were at a record level except for STT units: by mid-June they stood at an unremarkable (but by no means shabby) $3.20.

News came two weeks later that Brunswick had entered a full manufacturing licence which permitted it to incorporate OCP technology in Mercury outboard engines. Again, the market failed to react. As the final days of the sharemarket bull unfolded, STT units languished while the truly speculative issues on the national sharemarket commanded the attention of impatient investors.

Immediately before the global tidal wave of sharemarket selling hit Australia on 21 October 1987, STT units had eased to $3.15 each. When STT unit holders woke up on the fateful morning, their units were worth just $1.85. The crash had stripped $1.30 from the value of each unit. This meant that the paper value of STT had fallen from $928 to $324 million in a year.

Unfortunately, that was not the end of the bad news. By year's end, the units were selling for $1.10. Some finance commentators pointed out this represented a major bargain and they were right. The units recovered quickly to $2.70 each and those investors with enough confidence to buy enjoyed a 100 per cent gain in just a few weeks.

The post-crash recovery was assisted by the release of the manager's 1987-88 interim report which gave details of the tests done by Ford in Detroit. These were everything that unit holders could have wished. The fuel consumption target — set by Ford — was 32 miles per gallon (8.8 litres per 100 kilometres) but the OCP Escort achieved 36.8 mpg (7.7 l/100 km). The zero to 60 mph (0-100 km/h) acceleration target was 14.8 seconds but the OCP Escort took 11.2 seconds. The OCP Escort also exceeded three critical emissions targets and its emissions of nitrous oxide, the most dangerous of the exhaust pollutants, improved as the ambient temperatures dropped. In most engines, the NOx emissions rise as the temperature falls.

Directors of STM also announced that an Asian manufacturer (not specified but rumoured to be Hyundai) had signed a manufacturing agreement but that the agreement would not become operative until certain aspects of that manufacturer's operation came up to scratch. The Asian company, said the report, was confident that it would be the first to have an OCP engine in production and expected it to come off the production line in 1989.

STM directors also announced that the latest OCP engine produced NOx emissions at the rate of 0.24 grams per mile, a level well below the figure achieved in the OCP Escort tests where 0.69 grams was recorded (the target had been 0.80 grams). The latest NOx figures came from continuing development of the basic process and were under the proposed new standard of 0.4 grams being considered by the US Congress. At the same time, the OCP engine was 20 per cent more economical than the 'best in class' US engine.

STT units reached a peak of $3.42 early in March but investor confidence took a beating when the results of a joint Federal and WA government feasibility study were released. The report said that the manufacture of the OCP engine was feasible in Australia – provided that the development of the engine was successfully completed and a substantial international market was found.

Sarich was furious and OEC engineers protested that the engine had already been developed to the stage where it could be mass-produced.

The WA State government did not endorse the report but the Federal government defended it on the grounds that no comprehensive production engineering studies had been completed and no protracted on-road testing had been done by potential customers of the proposed engine plant. Sarich disagreed not with the study but with the manner it was released by Senator Button. He also later conceded the need for an extensive fleet-testing program.

Though disappointing to OEC, the study report did not stop STT units from rising to $4.08 by mid-June. Later that month, STM directors announced that Ford had signed an agreement permitting the Detroit giant to incorporate OCP technology in its engines worldwide. OEC had landed its first major car maker, although some technical and business details remained to be resolved.

By late June, the units had reached $4.50 again, despite a welter of inaccurate reports concerning the Ford licence agreement. The Australian press was still having trouble understanding the difference between the Orbital rotary engine and the OCP reciprocating unit. Some papers reported that Ford had agreed to use the old Orbital engine, the development of which had ceased four years earlier, while the country's major financial newspaper depicted the Ford arrangement as a research agreement and not a production agreement. Once again, the Perth Stock Exchange had to ask what was going on. In a supplementary statement to the exchange, STM directors said they had been 'stunned' by the newspaper report and went to the trouble of obtaining statements from Ford Australia and Ford US which underlined the importance of the agreement. It turned out that STM was right: no other agreement with Ford was necessary if Ford wanted to manufacture OCP engines in mass-production quantities.

The manager's statement correcting this obtuse reporting stopped the slide after STT units had dropped to $3.25 each by early June. They bounced back to $3.60 but a sudden drop to $2.80 on 2 August prompted yet another query from the Perth Stock Exchange. This time, the manager gave the bare-bones 'no, no, no' answer.

The price continued to drift during subsequent months and the situation was not helped by the change in the taxation status of unit trusts announced in May 1988 by the Federal Treasurer, Paul Keating. He declared that trusts would no longer be exempt from direct taxation and this meant that the principal advantage of using a trust structure to house the stake in OEC was nullified. The directors therefore proposed to switch to the more usual company status and adopt the name Sarich Technologies Ltd (STL).

Unit holders approved the proposal on 15 November, clearing the way for a renegotiation of the holdings of STL and BHP in the Sarich Technologies group to remove the limitations which could inhibit growth. The new scheme would make it possible for BHP to sell its share of OEC to Sarich Technologies Ltd in return for shares in STL. Such an arrangement would make BHP a major, but not the controlling, shareholder in STL. It also meant that STL would be the sole owner of OEC, effectively making it fully listed on the Stock Exchange.

The change in status was not expected to excite much investor interest and the price continued to ease towards the end of 1988. It finished the year just under $2 a share and continued into 1989 at that level. Despite all the success the company had achieved signing agreements with international engine makers, the company's value had fallen to $324 million.

Only time will show whether the market's assessment is right. Meanwhile, the history of the group's listing on the stock exchange shows how hard it can be to educate investors and the press in new ideas and technologies. It also shows how difficult it is for a stock exchange to control wild fluctuations in share prices regardless of whether they are the result of sky-blue optimism or misleading rumour.

In its short career on the Stock Exchange, the Sarich Trust plumbed the depths and, more often, the heights of Stock Exchange listings. The curious aspect of the whole affair is that the stock price had fallen dramatically at the end of 1988 just when the pay-off from two decades of furious work seemed to be around the corner.

Overleaf: The power and the passion. The Orbital Engine generated an enormous amount of press hype and inaccurate speculation during the early 1970s.

Page 129 Top: To attract interest in his unusual engine, Sarich made a number of personal appearances. Taken during 1972, this photograph shows that the four-stroke, seven-chamber unit was considerably more complicated than the press reported.

Page 129 Below: Sarich, Vic Brisboune and Ken Weaver prepare an Orbital engine for further testing. The engine was destined for a Cortina.

TWO SPORT SPECIALS

WHEN LARWOOD MET LILLEE — Page 3

DAYANA— THE HORSE THAT MONEY CAN'T BUY — Page 11

THE Sunday Telegraph
INCORPORATING THE SUNDAY AUSTRALIAN

WEATHER
Mostly dry
Details, back page

Vol. XXXV No. 1 SUNDAY, JANUARY 7, 1973 PRICE 10c*

Steel giant stuns motor industry

BHP TO BUILD PEOPLE'S CAR

AUSTRALIA'S largest company, Broken Hill Proprietary, plans to make a family car using a revolutionary Australian-designed motor.

BHP revealed this week that it is negotiating with a car manufacturer — believed to be Renault Australia Pty Ltd — to join forces to build the car.

BHP wants to become a partner in a venture which makes cars powered by the orbital engine and gearbox designed by 33-year-old Perth engineer Mr Ralph Sarich. BHP is financing the development of the engine.

The new car could look like this model, designed for Mr Sarich by Mr Wayne Draper, of Melbourne.

Engine for $60

Mr Sarich estimated his engine would cost $60 to make and $120 to retail.

The car will probably be a six-seater and Holden-size, designed for Mr Sarich by Wayne Draper, 25, of Melbourne.

The disclosure has stunned local car manufacturers who believed it would be a long time before BHP had plans for a new car. BHP announced only five weeks ago that they had bought rights to Mr Sarich's engine.

Although it is too early to set a price on the new car, it is expected to be inexpensive compared with similiar-sized models on the market.

Mr Draper said in Melbourne yesterday that he had already done several design exercises on which he hoped the new car would be based.

"With the Sarich engine being so small all you've got to do is fit the passengers in," he said. "Because of the engine's size we no longer have to design the car around the engine."

"Every car manufacturer in the world has been trying to get rights to the Sarich engine," Mr Draper said.

Technical details and dimensions of the new car, dubbed appropriately the Sarich, had yet to be worked out.

In Perth yesterday Mr Sarich said an agreement could be reached with the car manufacturer early next month.

CONTINUED PAGE 13

Mr Ralph Sarich

FREAK TEST RUN-OUT—SEE PAGE 48

Above: This publicity shot appeared when Dr Felix Wankel and NSU (the German car maker) announced a single-cylinder rotary engine in 1959. This prototype weighed 11 kg and developed 21 kW (29 bhp) causing pundits around the world to predict the concept would revolutionise motoring. Major car companies jointly spent over one billion dollars during the 1960s developing the Wankel engine for production vehicles.

Right: In 1973 GM said it would manufacture this Corvette as the first of a line of Wankel-powered cars. None went into production due to technical problems and the program fell in a heap after hundreds of millions of dollars had been spent.

Corvette
Four rotor

Above: Sarich used a different approach and called his engine an Orbital design because the rotor does not rotate about its own axis but follows an orbital path. The action of the sliding vanes can be seen here.

Right: The inside of the 3.5-litre Orbital engine. Note how the vanes act as partitions for the combustion chambers and slide in and out of the outer housing as the rotor turns.

XCP-326

Left: Sarich preferred testing the engine on a dynamometer where controlled conditions could be maintained but ran into flak from the motoring press which almost demanded to see the engine in a car. Here it is being installed in a Ford Cortina in late 1973. An Orbital engine was also fitted to a Renault but little road-testing was done.

Above: A Sarich technician, working in the engine bay of the Cortina, graphically demonstrates the Orbital's compact size.

Above: In May 1974, Sarich and BHP signed an agreement with Victa Ltd for the development and marketing of the Orbital engine for lawn-mowers. Victa designed a 209 mL (209 ccs) unit, 150 mm in diameter and 150 mm long. The unit was extensively tested but never put into production.

Below: During the 1980 Shell Mileage Marathon the winning OEC car achieved 2684.7 miles per gallon with Carol Darwin at the wheel.

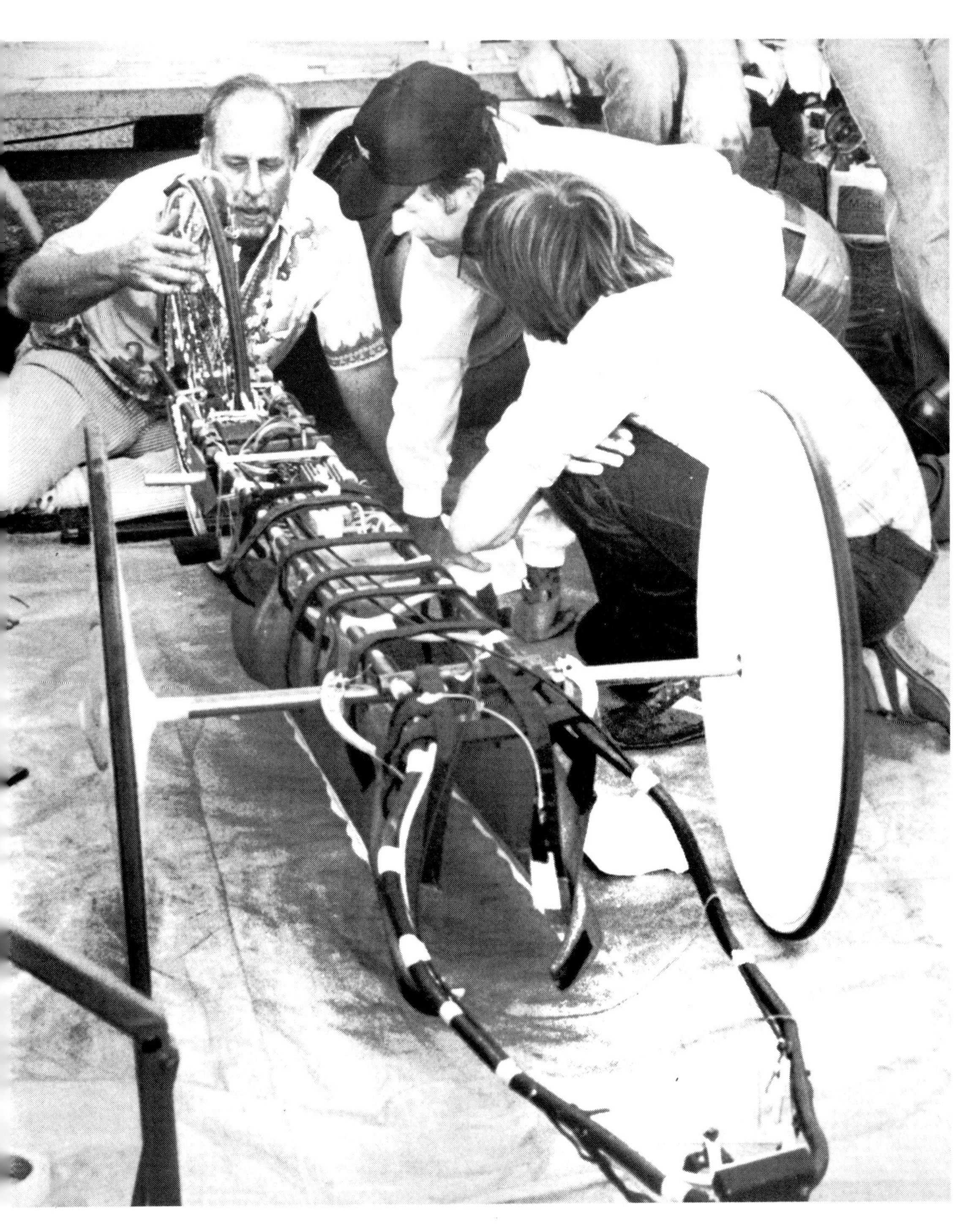

Above: Former World Champion Formula One racing driver Sir Jack Brabham is shown the secrets behind the record-breaking fuel-economy car developed privately by OEC personnel in 1980. Kim Schlunke is on his left.

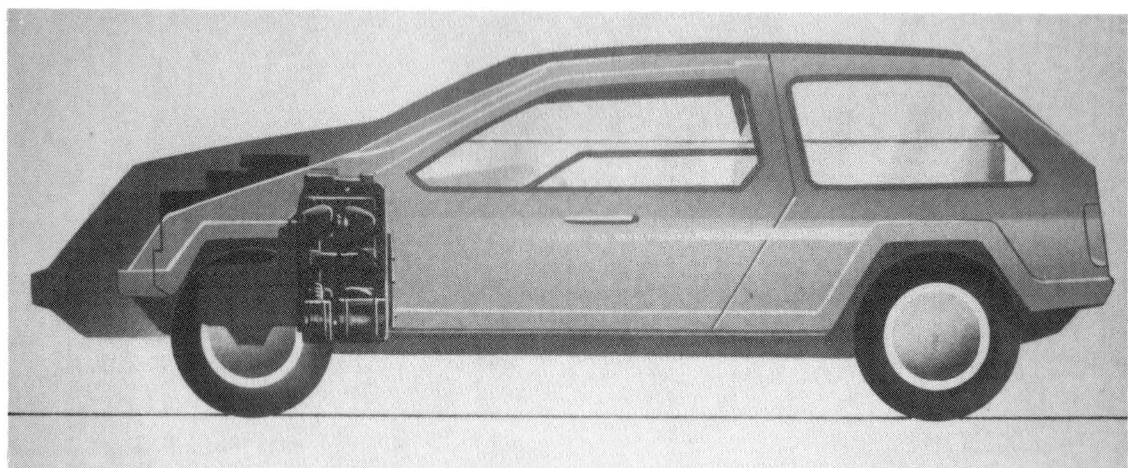

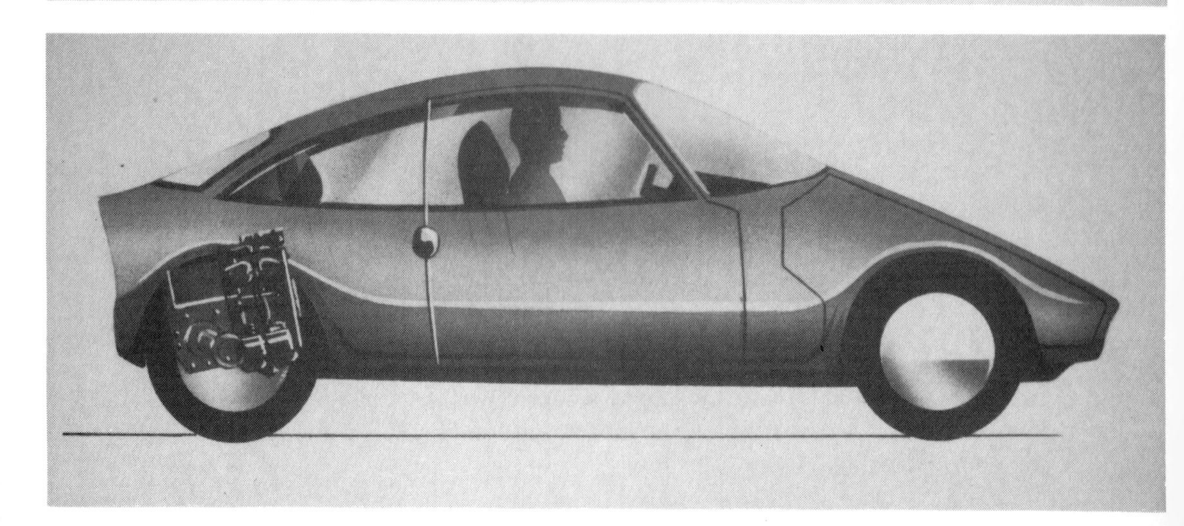

Left: The Orbital's main advantage was its compact size and light weight. Here the 3.5-litre version is compared with a conventional Rover 3.5-litre V8. The sketches below, from an OEC sales brochure, are intended to show how full advantage could be taken of its compact size.

Above: Orbital Engine Company's headquarters are in Balcatta, a suburb of Perth, Western Australia. Its facilities to test experimental engine processes are among the best in the world. All OCP components, including electronic gear, are made in-house.

Above: By the late 1970s, Sarich and his team had become increasingly involved in a novel combustion system involving a stratified combustion chamber and two-stream fuel-injection. This manifold version, known as OFIS (for Orbital Fuel Injection System), was fitted to a Holden Camira during the early 1980s. It dramatically reduced the fuel consumption and exhaust emission levels.

Right: Sarich prides himself on being a 'hands-on' managing director and can operate virtually any item of equipment in OEC's high-tech laboratory. (Sunday Times picture.)

Above: The emerging manifold fuel-injection system was tested in a six-cylinder Falcon around 1983.

Right: Sarich is smiling in April 1984 because, after 15 years of hard work, he can smell success. He is standing alongside a single-cylinder engine in which his new combustion system is being tested.

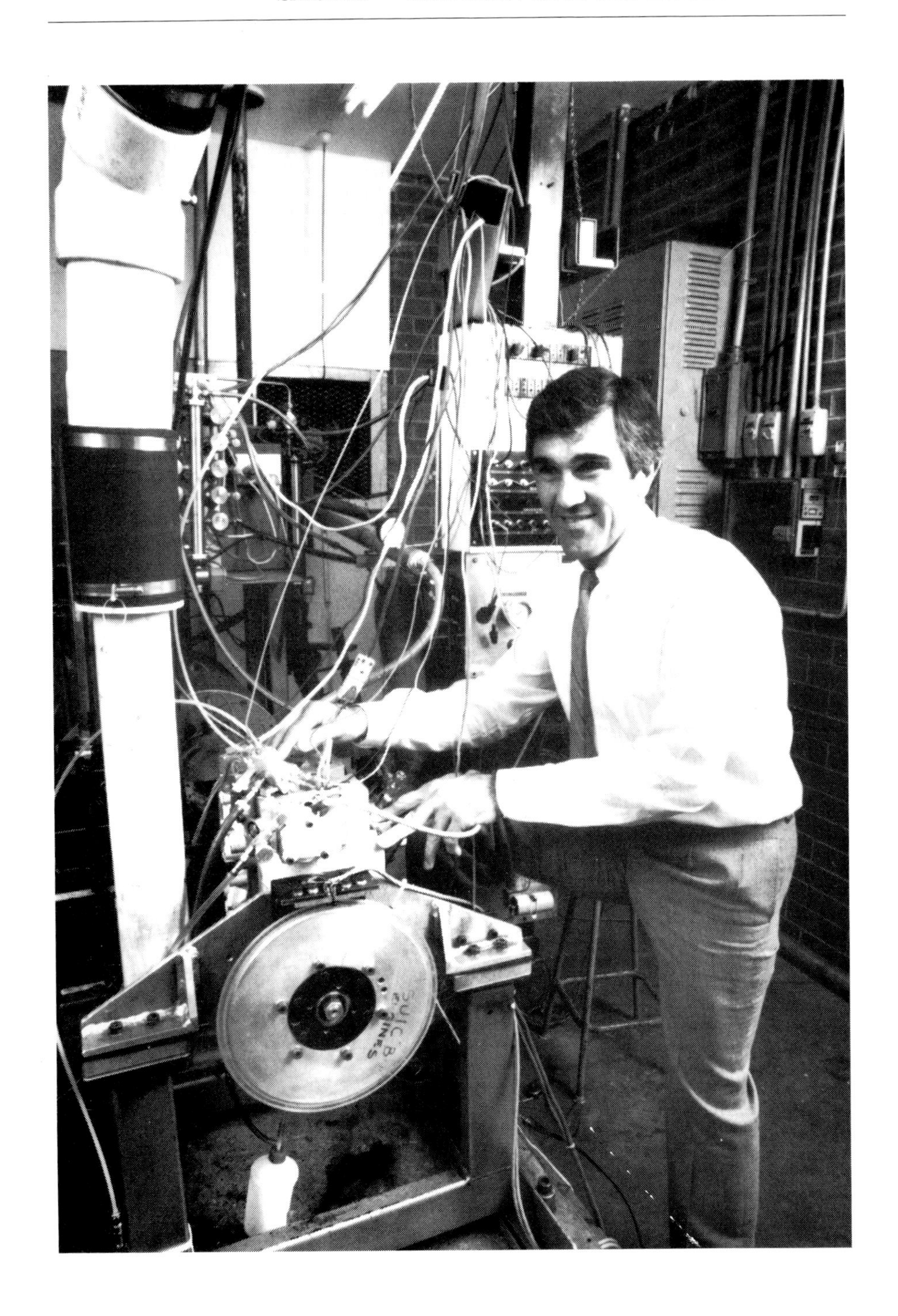

Above: The novel combustion process, which later developed into the OCP system, was initially tested on both Orbital and reciprocating engines. Here Ramsay Newmann is fitting a test rig to a single-cylinder reciprocating engine.

Above: Ralph Sarich with (left to right) Bob Thomas, Ken Johnsen and Kim Schlunke with the new OCP engine in May 1984. By this time the combustion system was showing tremendous promise.

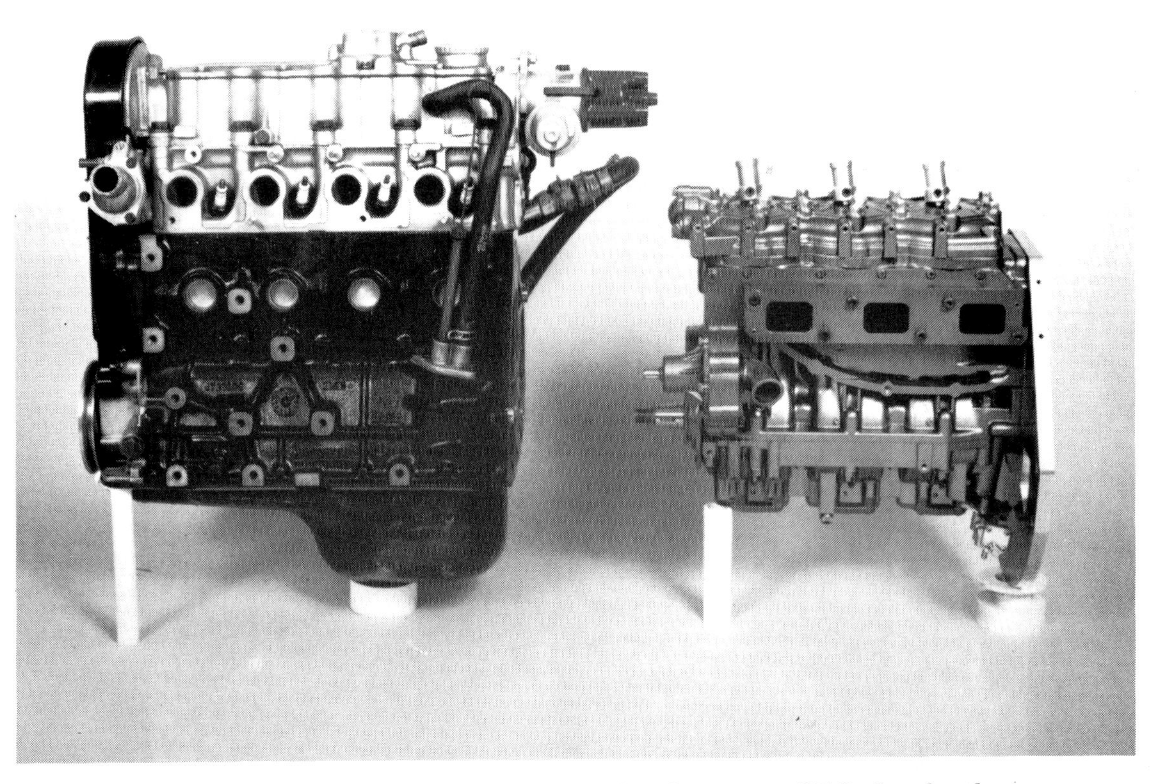

Above: To demonstrate its size and weight advantages, OEC placed a three-cylinder unit against the 1.6-litre Chevrolet Spectrum 'Family II' unit in August 1985. At the time, the smaller unit developed the same power but used less fuel and produced fewer exhaust emissions.

Left: In September 1984, Peter Czwienczevck works on the first two-cycle OCP engine installed in a car. This Suzuki-based C-series was fitted to a Holden Camira.

Above: Sarich jubilantly shows the press that the three-cylinder OEC engine (labelled 'A' on the display) has significantly fewer parts than a conventional four-cylinder unit of similar output (labelled 'B'). (Sunday Times picture)

Right: Don Railton demonstrates the remarkably low weight of the OCP, two-cycle reciprocating engine.

ORBITAL
COMPANY

7DT·733

Above: Ford Motor Company was the first car manufacturer to sign for world rights to Sarich's engine. Here an R-series model is installed in a Ford Escort for testing in Detroit.

Left Top: Tony Punch assembles an R-series engine for fitting in the Chevrolet Spectrum driven by author Pedr Davis.

Left Below: In June 1987, Pedr Davis became the first motoring journalist to drive a Sarich-powered car.

Overleaf Top: On 25 September 1987 a OCP-powered Escort left Perth for Detroit as part of a history-making agreement. For the first time, Ford will buy major engine technology, and probably complete engines, from an outside firm.

Overleaf Below: As built in 1987, the R-series weighed 65 kg (complete) and developed 63 kW (85 bhp) from 1.2 litres.

CITY OF ADELAIDE

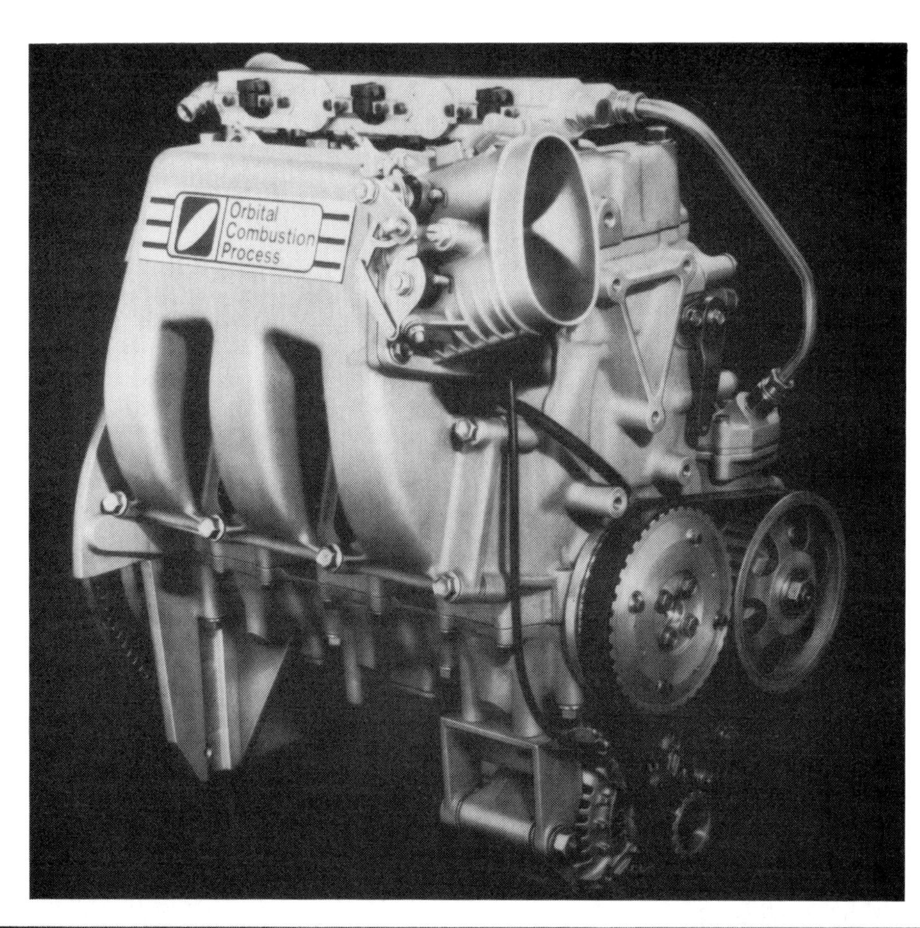
Orbital
Combustion
Process

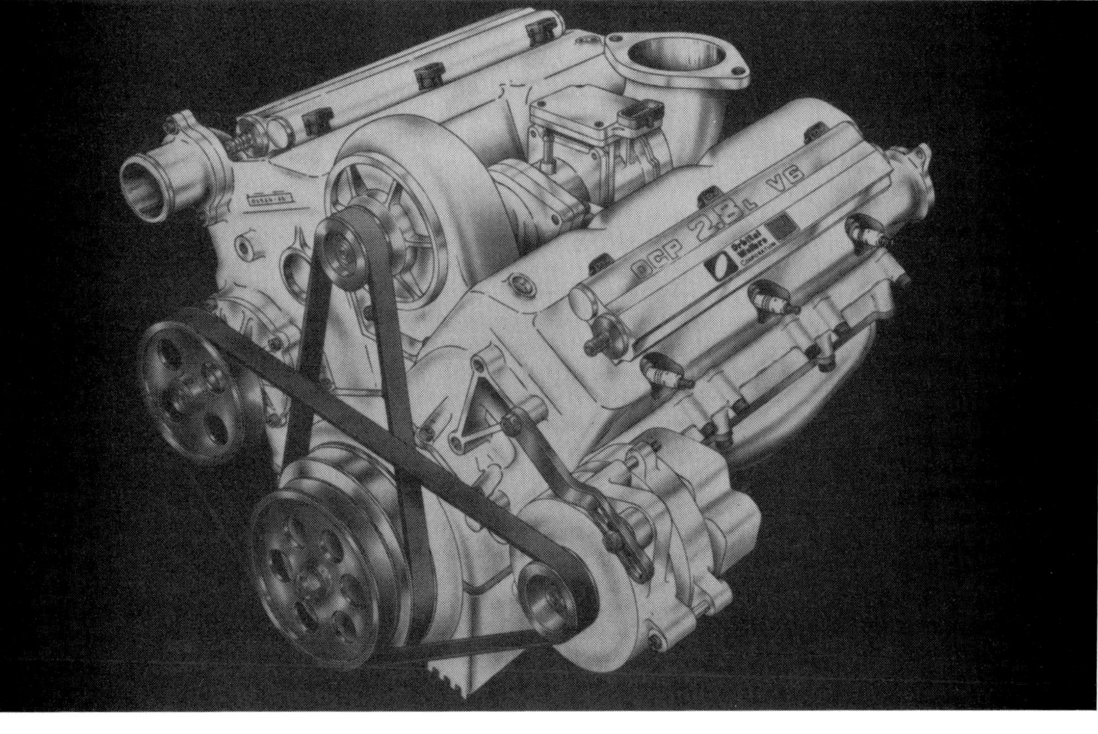
OCP 2.8L V6

Previous Page Top: By 1989, the X-series engine developed 78 kW (105 bhp) for a weight of 41 kg. Sarich had realised his 20-year-old dream of designing a lightweight engine more socially acceptable than the conventional type.

Previous Page Below: A supercharged V6 was under construction in early 1989. It is expected to develop in excess of 185 kW (250 bhp) from a capacity of 2.8 litres.

Above: Walbro Corporation has signed a joint agreement with OEC and BHP to produce injection hardware for use by the OCP engine manufacturing factory, various marine companies and possibly General Motors. The Orbital-Walbro factory is an important component in future OEC production plans. Founded in 1954 with its headquarters in Cass City, Michigan, Walbro makes a range of automotive and non-automotive fuel systems and has factories in several parts of the world. Company chairman Bert Althaver is seen here with the injector test equipment used in the development of Orbital-Walbro products. He holds a prototype fuel injector.

Left: As installed in an otherwise standard Hyundai Excel, the latest Sarich engine leaves enough room for a technician to stand in the engine bay.

Above: The small package size and the ease with which the engine can be installed is shown by this Hyundai installation.

Right Top: Ralph Sarich successfully launched his public company in May 1984. At one stage, the market valued the company at almost $930 million, though not a single production engine had been sold. The success of OEC has made Sarich one of Australia's richest men.

Right Below: One of the original Sarich partners, Tony Constantine is now chairman of Sarich Technologies Ltd.

Left: Ken Johnsen, manager of OEC, has worked for the company since joining Sarich as a 17-year-old trainee engineer.

Below: In July 1988, chief engineer Kim Schlunke was granted the Rolls-Royce-Qantas Engineering Award for Excellence for his 'role in the conception, development and marketing of the OCP system'.

Above: By 1988 the motoring press again expressed a keen interest in Sarich's operation. In March that year John Clydesdale of Perth's Sunday Times picked up an R-series OEC engine to show his readers that something special was afoot.

Below: 'You don't have to speed to have fun with a Porsche,' says Sarich, talking about his 928S4. (Sunday Times picture.)

Left Top: When the OCP engine is in production, family cars may look like this Ford Saguaro exhibition car, unveiled in 1988. The sloping bonnet suggests that a production version will have an extremely small engine, possibly an OCP V6.

Left Bottom: In October 1988, Ralph Sarich was awarded the prestigious Churchill Medal by The Society of Engineers, London. He was the first person outside Britain to receive the award, presented here by Winston Spencer Churchill, MP.

10

Manufacturing the engine

The typical automotive engine consists of several hundred individual parts, most of which involve a complicated manufacturing process. It is the most expensive mechanical component in the car.

A few specialist firms, such as Morgan and TVR, buy their engines from mass-production companies but most make their own, even firms such as Rolls-Royce and Aston Martin who sell fewer than 40 cars a week.

The commercial vehicle field is different and some engine specialist producers, such as Perkins and Cummins, build engines for a variety of rival truck makers. In such cases, price is not as high a priority as with car engines because trucks are often tailor-made to suit a specific purpose. Operating costs and financial returns are more critical than the initial price. But in the volume end of the car market, buyers are extremely price conscious and competition is extraordinarily fierce. In this environment, production costs are so important that saving even one dollar on a major component is an attractive goal.

This fact alone makes it extremely difficult for an outside supplier to profitably sell engines for use in mass-production cars. Even if a specialist producer could sell several hundred thousand units per year, it is unlikely that the producer could be price competitive. And, even supposing the price was right, a car firm already equipped to make engines would need a very good reason to switch to an outside supplier.

High numbers are the name of the game and some firms think in telephone digits. For example, Ford built 949,000 Escorts in 1987, Toyota sold 674,215 Corollas and 349,814 Camrys. An enormous investment in robotised and automatic production machinery is needed to produce engines in these numbers, but once installed, the equipment cuts the price per component to an almost irreducible minimum.

The only way to crack into the market is to produce an engine which is markedly superior to existing designs or one which could be manufactured substantially more cheaply than the conventional kind.

Low manufacturing costs are not strictly a matter of numbers. After a certain production number has been reached, there's a cut-off point after which no further cost advantages can be gained by raising production.

In recent years, studies have shown that, with the latest technology, a production run of 250,000 engines per year is ideal. A higher figure introduces complications without necessarily saving money. A lower figure raises the unit production cost to the point where the engine is too expensive for the price-competitive segments of the market.

This figure of 250,000 units has become the 'industry module' because it is the smallest number required to support the installation of the latest foundry and machining equipment and produce a cost-effective product. It serves as an industry costing standard now widely used for comparison purposes.

'Of course, 250,000 engines a year is not an inviolate number', says OEC's Kim Schlunke.

'It is just the economic number for any one plant.

'Some firms talk in terms of 400,000 units per module, but GM has modules of 250,000 for the Family II engine which is produced in Austria, Spain, Brazil and Australia. They have two or maybe three such modules in Austria alone.

'The engines produced in one module do not need to be identical, although they must come from the same family for economic reasons. In our case, there's plenty of room for variations within the OEC family, especially as we can combine two three-cylinder blocks to form a V6.'

Mass-production plants are expensive. In the early 1980s, the industry had a rule of thumb which held that the cost of establishing a completely new plant amounted to a million dollars for each thousand engines it could produce per annum. In other words, a plant with the capacity to make 250,000 engines a year was expected to cost US$250 million including the land, factory, machinery, assembly line and all-important foundry.

Prices have risen sharply since that rule of thumb was established and $400 million is considered closer to the mark for 1989. This figure applies to a conventional four-cylinder engine with the normal valve train required for a four-cycle operation. Because of the reduced number of components, the capital outlay needed to produce the three-cylinder OCP unit would be appreciably less than for an equivalent four-cylinder, overhead camshaft design.

'In conjunction with the Federal government, GMH and Ford, we did a costing exercise for a plant to produce our engine and came up with a figure of A$290 million in capital outlay', says Kim Schlunke.

'This means that a factory for our engine would cost about $120 million less than a similar plant needed for a conventional four-cylinder, four-cycle unit. Our engine is basically less complex and the plant would not need the usual lines to machine and assemble the complex cylinder head and valve gear. In addition to the reduced outlay, the OCP unit would be about A$300 cheaper per unit to manufacture because it has fewer parts and less weight.

'I'll explore that $300 a little. In manufacturing terms, each engine has several prices. One is the selling price which includes development and profit, another the fully-accounted cost which takes overheads, but not development costs and profit, into account. And there's the internal cost. The fully-accounted cost is probably a better way to look realistically at the subject because it represents the total manufacturing cost.

'In 1988 dollars, it takes about $600 to produce a low-technology, four-cylinder, OHV, cast-iron, carburettor engine in Australia. We believe we can manufacture an engine with an equivalent power output for $400, a saving of $200. These low-technology units will not be in production for much longer, however, because of the stricter emission laws coming in around the world.

'It's best, therefore, to compare us with a current US four-cylinder unit with electronic fuel-injection. That would cost between $700 and $800 to produce and this means we would have a $300 advantage.

'This sort of number has a very significant effect on the car's ultimate retail price and its competitive position in the marketplace.

'Because we have a performance and emission advantage over any production engine of similar capacity, $300 is a valid cost advantage to quote. It assumes a comparison between the OCP engine and a modern US engine capable of meeting the US emission standards.

'There is no point in comparing us with an engine which cannot meet the emission standards.'

These cost claims are based on volume production and, in this case, the target production is 250,000 units a year. There's no doubt that the OCP is slated for production, but as of late 1988, neither Ralph Sarich nor Kim Schlunke was prepared to provide a timetable. But piecing together private and public information, the scenario is likely to be as follows: towards the end of the first quarter of 1989, Orbital Engine Company will combine with the Australian government and possibly an Australian and a US State government to announce financial support to develop the engine to the 'signing-off stage'.

This industry term means subjecting the design to an extensive test program in which engines are installed in a fleet of cars and tested for durability and possible unforeseen problems. Only when the OEC engineers and potential customers are satisfied that all bugs are out of the design will it be 'signed off'. It will then be ready for production without further evaluation or changes.

Fleet testing is very expensive because it means that a sizeable number of engines have to be virtually hand-built, in this case at the rate of one unit per day. Engines are extremely expensive to build by hand and each could cost in excess of $20,000. These engines will be fitted to perhaps 20 vehicles which are driven by a variety of drivers under a range of operating conditions for six months or more.

In addition to durability, the field testing is designed to flush out unforeseen problems which could arise in normal use, such as idling in a lengthy traffic jam on a hot day or metallurgical difficulties associated with high-speed driving. Although all car designers attempt to anticipate every conceivable form of abuse during the development period, actual driving in everyday situations can reveal a potential problem no one had thought about. A few years ago, one Australian manufacturer ran into the embarrassing situation where its engines did strange things when a police radio was used nearby. The signal from the radio transmitter seriously interfered with the engine's management system and could even change gear in the computer-controlled transmission. A major importer ran into trouble when dealers fitted a particular brand of air-conditioning which interferred with the standard cooling system, causing the radiator to boil.

To carry the high cost of 'signing off' the OCP engine, a consortium of interested parties has been put together and — it is the author's guess — this includes Ford, a US two-cycle engine maker and a Japanese company. Although these firms are customers for the final product and are pushing to have it in production as soon as possible, they will not necessarily be partners in the manufacturing venture.

To fund this $300 million venture, separate negotiations are underway. The partners will be carefully chosen because they will be expected to provide expertise as well as cash. If possible, one partner will have experience in robotic production techniques, one in high-volume automotive engine production and another with two-cycle engines.

'Ideally, we should have some kind of financial institution as well', says Kim Schlunke.

'The debt equity ratio in our production plant will probably be in the order of two to one; that is, the partners would put in about $100 million in cash and the rest will be borrowed. This being the case, the consortium would benefit by having a financial specialist.'

Ralph Sarich stresses that an experienced maker of volume-produced engines must be involved, but as of early 1989, the big question was not if the engine would be mass-produced — but in what country.

'You could produce it almost anywhere, provided you have ready access to experienced and well-educated labour', says Schlunke.

'Our studies show that the transport cost of bringing in individual components or shipping out complete engines is fairly small when factored into the whole equation.

'In this sense, we could build the plant in Western Australia, Victoria, South Australia, North America, Asia or Europe. I don't see freight as the main problem.

'Ideally, we should build the engine in Australia, but the difficulty is that it is more expensive to produce here than elsewhere.

'There are several reasons. Australian workers are on the job for 219 days per year whereas North American firms have something like 245 working days a year. This important productivity difference between Australia and the USA worries the potential partners in the consortium.

'Another consideration is the supply of components such as stampings, pistons and bearings which traditionally come from outside. No matter how good the Australian supplier may be, he is geared to service car makers who typically produce volumes of 60,000 units per year. The supplier is locked into that kind of volume and the cost of any given part is more expensive than you would expect for an identical part bought from a typical American producer.

'These cost drawbacks are the main reasons why Australia is a less attractive place to produce engines than the USA. And remember, there is no way we could sell 250,000 engines a year in this country. We have to sell on the world market, competing against engines built in countries as diverse as Taiwan and Japan.

'For all that, a detailed study shows that it could be commercially viable to produce the OCP engine in Australia provided that the equity holders were prepared to sacrifice some profit.

'As OEC is likely to have only a 30 per cent equity, Ralph has the difficult job of convincing our partners that the engine should be made here.

'In order to convince them, Ralph has been arguing that someone should make up the deficit. This call has been misconstrued by the media. He is not asking for a handout or saying the government should fund a private venture. He is asking that the Australian government considers the matter as an overall business proposition. We believe it makes good sense for the government to pay a subsidy to keep production here.

'In my view, it would mean some obvious economic and social advantages, such as bringing extra employment to the area where the factory and suppliers are located and attracting export earnings amounting to over a hundred million dollars per year.

'If the government agrees to a subsidy in the order of $30 million a year, it will find it is making money on the deal when it balances the books at the end of the day.

'The decision where to locate the engine plant is not just ours to make on patriotic grounds. We will have overseas partners who will look at the position purely from a business point of view.

'We want to persuade them that the plant should be in Australia. On the other hand, when the government balances its books, we want to be sure the public purse is not making a loss. That would constitute a handout and that's not what we want.

'Having said all that, I have a strong feeling that the project should come here despite the need for a subsidy. The government has already spent half a million dollars to find out that the OCP engine is a viable project for Australia. Our problem is that the equity holders in the consortium would not make as much profit here as they would by making the engine in the USA.

OEC is confident that it can keep a plant busy almost from the day it opens. Says Schlunke:

'Obviously we would not be going ahead with the program unless we had a fairly firm commitment from at least one major car maker and we have one. There's a definite market for 250,000 OCP engines a year from day one.

'I expect several more customers. Technically we are a long way ahead of the US emission control legislation. There is no published data anywhere which suggests that anyone has achieved 44,000 mile emission levels [a US requirement to ensure the system maintains a clean exhaust] as good as ours.

'It's a pretty good shot that we shall see a car on the road with an OCP engine during late 1992.'

Who will make the car?

'Although Ford was first to sign a licensing agreement, it does not follow that Ford or even GM will be first in the market place with an OCP engine. We are dealing with a Japanese firm noted for its ability to get new products into the showroom remarkably quickly', Schlunke says.

'It's an interesting fact that all our customers have written their specifications for the new US emissions standard, so maybe the engine will be used first in the USA and not Europe as most people had predicted', adds Ralph Sarich.

Although the OCP engine itself is almost ready for production, a considerable amount of painstaking work still needs to be done. OEC has its sights set on commencing small volume production in 1990 when the first pilot production run will be made using automated equipment. Some machinery needed for pilot production will be used a year or so later when volume manufacturing commences. Logically, the pilot and volume production runs should be made in the same place, hence the need for an urgent decision on the plant's location.

The pilot run could be as high as 500 engines per customer and this pushes the pre-production and testing costs well into the big league — possibly well above $100 million. There is no way to do the job cheaply. No one — OEC or its customers — could afford to market anything less than a fully sorted and proven engine.

Even the signing-off program, without the pilot production run, is likely to cost around $40 million and this sum provides 15 to 25 cars and engines for fleet testing. The details of both programs were being finalised during early 1989.

'There is an element of risk that today's plans won't come together', said Ralph Sarich in December when discussing the forward plans.

'But we are having intense pressure applied from the car companies to get production moving. We had been stalling hoping that the Australian scene will improve, because this is where I would prefer the production facility to be.

'The main drawback here is the US$30 million dollar net value difference per year between manufacturing the engine here and in the USA. Britain has also made a proposal and we've had offers from Ireland and Singapore.

'But the likely scenario is that we should have everything in position by the first quarter of 1989 and an announcement will follow soon afterwards. It will possibly be made jointly by a US state governor and the Australian government. This will be the first time in history that two governments have combined to launch new technology.

'It is being increasingly recognised that there's immense social value in the engine because of its low level of exhaust emissions and even now the US Congress is looking at us and the emission numbers the OCP engine can produce.

'This is one reason why a US state government is willing to invest in the engine plant even if it is built in Australia. And possibly vice versa. In the longer term, I hope there will be one plant over there and another here.

'If the manufacturing project achieves the estimated levels of business, we expect between one and two million of the current design will be made and some versions could sell for an average price of $1000. This would mean enormous earnings for the country producing the engine.'

The state of Michigan — and at least one other US state — has a strong interest in the project. Serious discussions started in 1986, and in May 1987, Ralph and Pat Sarich visited the state as guests of the Michigan state governor, James Blanchard. In March 1988, five of his top-ranking officials flew to Perth to present an incentive proposal they hoped would secure the engine plant for the USA.

Meanwhile, in June 1987, the Australian Federal government and the Western Australia State government formed a project team to investigate the manufacturing potential. The team comprised key people from Holden's Engine Company, Ford Australia, BHP, the Australian Council of Trade Unions and OEC.

The study cost about $500,000. The final report of the steering committee, issued in March 1988, concluded that Australia has the technical and resources capacity to produce the engine. It also confirmed that a single plant producing 250,000 OCP units a year would be financially viable and would provide a significant benefit to the Australian economy.

The government of Western Australia has worked hard to keep the project in Perth where parochial feeling runs high. In early 1988, the Western Australia Liberal Party joined the State Labour government in an approach to try to persuade the Federal government that the engine plant must be built in Western Australia.

'The plant is too important to WA for it to become a political football', said the State's leader of the opposition.

'We should be making a bipartisan approach to Canberra for the necessary assistance to guarantee that this plant is established in this State.'

Company manager Ken Johnsen believes the ultimate decision must be a business one.

'We have done our sums, and although the project would be viable in Australia, it would be much more viable in North America', he says.

'There's been a lot of nonsense published about Ralph trying to force the Australian government to pay for the production plant. The people who write this rubbish don't realise how genuinely he wants to get this motor in production for the overall good of everyone. He is committed to producing an engine which will permanently benefit the community. He is not looking to line his pockets with a few more million dollars.

Money is not his motivation. He is not like those multimillionaires who treat success by the amount of money they make.

'I personally find it most annoying when people say: "Why should the government help Sarich? Surely he's got enough money to fund this engine plant?" They do not understand the size of the financial undertaking nor that he is putting his personal fortune more at risk than the Federal government or any State government.'

By early 1989, when this chapter was written, all signs pointed to 'go'.

'We are in the process of finalising the joint venture partners in the manufacturing company', said Schlunke.

'We have one firm customer and a number of prospective customers all of whom are keen for us to get into production quickly.

'The sorts of people which come to mind as potential partners are American and Japanese marine companies because of their experience in two-cycle outboards. Some marine firms also have automotive experience. Mercury, for example, makes the new Corvette engine for Chevrolet and Yamaha is involved with Toyota in car engines. Outboard Marine also has experience in the auto field, so there is no shortage of suitable partners. It is largely a matter of negotiating and finalising a deal.

'We plan to put together a consortium and – what people seem to miss – the consortium will buy a licence from OEC and be an independent operation. We will, of course, have an interest in it, probably around 30 per cent. The consortium will pay OEC a licensing fee to produce the engine, but that fee may be waived in exchange for our equity in the manufacturing company.'

Whilst the discussions take place, there is plenty of action around the OEC headquarters in Balcatta, near Perth, confirming that a major expansion plan is underway, regardless of the outcome of the talks.

Ralph Sarich and company chairman Tony Constantine had jointly purchased a large parcel of land near the current OEC facility to cater for future expansion. They have now decided that the factory extensions should be closer to head office and OEC has leased the land immediately across the road. Here a modern, special-purpose facility is being built. It is not intended as a manufacturing base but to house additional research facilities, including those associated with the 'signing-off' and fleet testing programs.

OEC is also building distance accumulation facilities on another block of land adjacent to its headquarters. This is where vehicles which are to be certified for US emission compliance can accumulate test mileages on special dynamometers known in the business as rolling roads. Each prototype will be driven for 50,000 miles (80,000 km) with an electronically-controlled automatic driver putting it through an approved driving cycle which simulates actual usage.

In addition to the engines which will later be made by the consortium, at least one major firm, General Motors, will probably mass-produce its own versions under licence. At least one and possibly up to five Japanese firms are expected to manufacture OCP engines.

When the lower production cost is added to the unquestioned technical advantages arising from the low emission levels, light weight and greatly reduced packaging size, the OCP unit becomes extremely attractive even to car makers who already build their own engines in very large numbers. This is why Ralph Sarich is likely to become the first man since World War Two to sell a mass-produced engine to the major car makers. If the engine sells in the numbers expected, it will yield rewards of mind-boggling proportions.

Walbro Corporation

The joint venture between Walbro Corporation and OEC was established to produce electronic fuel-injection systems with key technicians from both companies providing the necessary skills.

Walbro Corporation designs, produces and markets a wide range of precision-engineered fuel systems for the automotive and small engine markets. The company manufactures fuel delivery subsystems as well as components for automotive fuel-injection. It makes electric fuel pumps for the automotive aftermarket and the agricultural, industrial and marine markets. Its carburettors are used with commercial and consumer chainsaws, string trimmers, lawn and garden equipment, outboard marine engines and industrial and recreational engines.

The corporation has its headquarters in Cass City, Michigan. Its engineering and development facilities are located in the USA, Japan and West Germany. Manufacturing operations are located in the USA, Japan, Singapore and Mexico. Walbro is likely to be the principal supplier of the injection system to the OEC manufacturing venture, various marine engine companies and some automotive companies.

Walbro is easily the largest firm in Cass City, a small town where one in five of the 2250 residents work at Walbro and many are shareholders.

The company was founded in 1954 by Walter E. Walpole and initially funded by local businessmen to the tune of $25,000. Today it is a public company with a 1988 turnover of US$135 million; more than half its income comes from sales to the worldwide automotive industry.

In late 1987, Walbro was the target of an unwanted takeover attempt by a New York conglomerate, UIS Inc. UIS quietly purchased 6.8 per cent of the Walbro stock, then revealed its bid and tendered for a further 60 per cent of the 3.5 million outstanding shares. The offer of US$27.25 per share put a value on the business of almost $100 million.

Rumours swept through town that, if the company was sold, it would probably be relocated. Horrified at the thought, the directors, management and towns people, led by company chairman Bert Althaver, put up an almighty battle. Althaver wrote to all stockholders pointing out the inadequacy of the offer in light of the company's potential and asked them to keep faith. Some locals went into debt to buy more shares to help keep the raiders out. Many signed a pledge that they would not part with the shares they already had.

The management hired a legal firm and an investment banker specialising in takeover defences, cancelled trips and took what amounted to a crash course in corporate raiding.

'I did nothing else for eight weeks', Althaver told a reporter from Automotive News, Detroit's main industry newspaper.

Althaver also contacted GE Credit Corp who agreed to put up $35 million for a new issue of preferred shares. Walbro used this money to buy back 20 per cent of the outstanding common shares.

UIS Inc. retaliated with a lawsuit which sought to prevent Walbro from selling the preferred stock to GECC or buying its own shares. When the court refused to stop Walbro, UIS sought to disengage from the takeover. The fight was over.

The battle had been costly, with estimates of Walbro's costs ranging around $5 million, but at least the town kept its largest employer. The company shares later rose to almost US$30 each. Although Walbro denies it, the battle must have affected the timetable for producing injection systems for the OEC engine.

Even so, Walbro's involvement in the joint venture company, Orbital Walbro Corporation, puts it in fine shape to benefit from OEC's agreements with Ford and General Motors. It will also be the main supplier of low-cost injection systems to the OEC joint venture factory.

11

Technicalities — the Orbital engine

The concept of the reciprocating piston engine dates to 1711 when two Englishmen, Thomas Newcomen and John Calley, built an external combustion device which used steam as the working medium. The piston engine was therefore a practical reality long before 1770 when the first motor vehicle, the Cugnot steam carriage, made its appearance in Paris.

The modern petrol engine probably dates to the mid-1770s when Isaac de Rivaz (a Swiss citizen living in Paris) learned that a scientist named Volta was experimenting with the electrically induced ignition of gases. Realising the significance, de Rivaz applied the process to a single-cylinder stationary engine and thus became the first man to use electric ignition for an internal combustion engine. He patented his device in 1780 and later sketched out ways in which it could be used to power a vehicle. Contemporary reports said that a powered car, resembling the de Rivaz sketches, made a number of test drives in the area of Vevey on Lake Geneva in 1813.

Next man on the scene was Joseph Etienne Lenoir. Born in Luxembourg, he came to Paris in 1838 where he took out patents for an assortment of technical inventions including a teleprinter and an electrical brake system. In 1860, Lenoir was granted a patent for an engine in which the air 'is expanded as a result of combustion'. His power plant was smaller and cheaper than the fashionable steam engine and delivered as much power. It was widely acclaimed and the Gautier company was founded in 1860 to manufacture the new engine. More than 400 units were sold within a few years. Lenoir himself fitted one in a boat in 1861 and built a primitive 'horseless' dogcart a year later.

In terms of fuel efficiency and power-to-weight ratio, however, the Lenoir design left much to be desired. This fact was demonstrated at the Paris Exhibition of 1869 by means of a direct comparison between it and a new engine built by the German inventor, Dr Nikolaus August Otto. Although historians argue about who was first to build an efficient petrol-powered engine, it is certain that, in 1876, Otto patented the grandfather of the engine now used in most motor vehicles. In short, he devised the four-cycle principle used today and often called the Otto cycle.

It is worth noting, if only for historical reasons, that a motor vehicle powered by a four-stroke reciprocating piston engine had been built by Siegfried Marcus, an Austrian, six years earlier. The Marcus 1870 engine ran on similar principles to the Otto design, although it was Otto who secured the patents.

Otto's four strokes, or cycles, are, of course, fundamental to the operation of a reciprocating piston engine. They are as follows: induction of the fuel and air, compression of the air/fuel mixture, ignition and subsequent expansion of the mixture and elimination of the spent gases. In the four-cycle engine, these actions take place during two complete revolutions of the crankshaft during which time each piston makes four movements up or down the cylinder. A two-cycle engine has the same four basic operations but they take place during one revolution of the crankshaft.

Until the Sarich combustion technology was demonstrated, it was generally accepted in automotive applications that four-cycle engines produce fewer emission problems, have better fuel economy and provide more low-speed torque than two-cycle engines of equivalent capacity. Two-cycle units, conventional wisdom believes, are not as easy to start when hot, less complex, lighter, less bulky and potentially more efficient. They also have less internal friction, need fewer volts for cranking and are generally easy to start when cold.

Despite the expenditure of vast sums on research, no one has yet produced an engine which seriously challenges the reciprocating piston unit as the most practical form of motive power available for a mass-produced vehicle. Certainly, many claims have been made for such unorthodox designs as the gas turbine, Wankel rotary, Stirling external combustion unit and various forms of reciprocating and rotary steam designs. Vehicles using these power plants have been successfully built but only the Wankel, as fitted to some Mazda models, has moved into mass production.

Despite its well-publicised faults, it is hard to quarrel with the proposition that the piston engine has the best overall combination of attributes for the job expected of it. It will not, therefore, become obsolete until someone produces an outstanding unit which is clearly superior to the piston engine in some major respects and at least equal in all others. Even if this is achieved, the newcomer will make slow progress towards volume production. The automotive industry is inherently conservative and reluctant to depart from a proven success story.

Too many firms have gone bankrupt trying to produce cars which were 'ahead of their time'. For example, NSU – which developed the Wankel engine and produced the first rotary-powered car – was in perilous financial position when Volkswagen took over the factory. Time and time again, events have shown that the average motorist is not interested in investing money in a vehicle which is seen as experimental or radical.

Like countless thousands before him, Ralph Sarich could see the need for a car engine which is lighter, cheaper, smaller and less costly than the traditional piston design. He started with a non-reciprocating unit, then moved into a far more promising area by making a radical improvement on existing piston engine technology. Unlike most other inventors, Sarich had the stamina, determination and financial support to persevere to the point where he demonstrated what is unquestionably a better mousetrap. In essence, he has an improved form of combustion technology – and car makers have beaten a path to his door.

Only time will show if the Sarich approach will become universal for the auto industry – but it certainly stands a far better chance than any major departure from conventional practice since the adoption of electronic fuel-injection.

As earlier chapters in this book have shown, Sarich made two distinct stabs at creating the smaller, lighter, more powerful engine he sought. The first was a type of rotary design which he named Orbital. It is still possible (but not probable) that it will see production. The second stab came as a spin-off from work done on the Orbital engine and took the form of a new combustion process to make reciprocating engines operate more efficiently.

By combining this combustion process with the known characteristics of the two-cycle piston engine, Sarich was able to produce a power plant which is demonstrably lighter, less costly to make, smaller and more efficient than any current production engine of similar output. It was this ability to improve the 130-year-old reciprocating engine which has earned him worldwide fame.

This chapter, however, is concerned with the technology involved in the original orbiting unit.

Background and development

The first rotary engine was developed from an idea by an engineer named Ramelli who, in 1588, built a water pump operating on the rotary principle. Later, in 1629, came a steam jet and, in 1772, James Watt devised a rotary gas turbine engine.

The first automotive rotary engine came in 1900 when Ljungstrom built a steam unit with a rotary piston. After that came hundreds of novel designs and thousands of variations on them, resulting in a large number of petrol-powered and steam-powered prototype engines. Some ran on two-cycle principles, others four-cycle, and most proved to be smoother and less bulky than existing piston engines.

In many cases the operating concept was relatively simple, but the execution was complex, creating some engineering difficulties which, so far, no one has completely solved. The basic idea in a rotary engine is to have one or more rotors (or pistons) spinning continuously in the same direction. As it rotates, the rotor creates one or more working chambers which vary in size as the engine proceeds through its cycle. Varying the size of the working chamber allows the normal Otto cycle – intake, compression, expansion and exhaust – to take place.

Numerous inventors have come up with ingenious geometries which provide variable working chambers using a minimum of moving parts, but most have run into technical problems.

One major difficulty has been in sealing the working chamber to keep the hot gases inside. Even when effective sealing has been achieved, the seals have been found to wear at an excessive rate when the engine spins rapidly, especially if it is difficult to lubricate them. Almost rivalling this as a challenge, have been difficulties surrounding the actual shape of the working chamber. Ideally, a combustion chamber should be spherical so that the surface area to volume ratio is minimised. This cuts down heat transfer and the production of hydrocarbon (HC) emissions and minimises the flame travel to reduce detonation and combustion duration. Although a perfectly spherical shape is virtually impracticable, the closer the designer can get to it, the lower the fuel consumption and emission levels likely to be obtained.

In a well-designed piston engine, a combustion chamber close to a hemisphere is normally obtained, but few rotaries have come close to this shape. The problems associated with obtaining a satisfactory shape for the working chamber as the rotor spins have taxed even the most ingenious geometricians.

Despite the difficulties, the advantages of rotary engines make them extremely attractive from an engineering point of view. Most offer the prospect of a lighter weight and smaller number of moving parts than a reciprocating unit. They are also compact in size and easy to lubricate.

These virtues were foremost in Ralph Sarich's mind when he started work on his unit. Sarich was not the first to have the rotor spin with an orbital or eccentric motion but he made the idea well known by incorporating this action in the engine's name.

His design differed from other orbital types in the manner in which the working chamber was sealed. Sarich deliberately chose geometry which caused the rotary piston to move in an orbiting motion (rather than about its own axis) because this reduces the number of problems associated with providing variable size working chambers. The success he achieved with this design came largely through the ingenious and effective seals he was able to devise.

The first Sarich engine had six combustion chambers and a nominal capacity of 2.9 litres and ran with a two-cycle action. It used direct fuel-injection with compressed air carrying the fuel into the combustion chamber.

'It ran only in spasms', he said later. 'The trouble was the poor sealing. Burning gas would leak back into the intake manifold and the manifold would catch fire. We converted the design into a four-cycle system with redesigned seals and a Holden carburettor.

'That's when we had some success and were able to show the engine running on 'The Inventors' show.

'Our next design had disc valves and seven chambers. We went to seven chambers for the same reason that aircraft engine designers use seven cylinders – you need an uneven number to get the timing right. That engine ran very sweetly but we had a lot of trouble sealing the disc valves.

'The discs were driven by one of the eccentrics and made the engine very compact. The system had several problems, however, including a high level of friction and we found it hard to make the valves stable. Although we had that engine running well, we learned, like everyone else, that it is very hard to make disc valves work without distortion.

'In parallel with this seven-chamber unit, we built a small engine with three chambers. It could rev to high speeds – around 8000 rpm I think – compared with 4500 to 5000 rpm for the larger engines. Even at these speeds, the small engine did not develop as much power as I had expected and I changed the design to four chambers.'

No tests were done with a diesel variant, though this was a practical option, and most work was on four-cycle petrol units. By 1980 the firm was concentrating on a 3.5-litre, seven-chamber design which showed great promise and came close to the point where volume production was a genuine possibility.

From the very start, Sarich used the name Orbital engine and described it as an internal combustion unit operating on Otto four-cycle principles. The final version, developed between 1980 and 1982, is approximately one-third the size and weight of a conventional six-cylinder piston engine with a similar displacement.

Basically, it consists of an outer housing (approximately cylindrical) and two end plates. A rotor (or piston), located within the outer housing and mounted upon a crankshaft, rotates in an orbital fashion guided by three (later four) specially shaped cranks known as eccentrics. The drum-like rotor engages with seven sets of vanes (with seals) which act as partitions and create seven working or combustion chambers. As the rotor spins within the housing, any one point on its periphery moves closer to and away from the housing because of the orbital motion. This, of course, changes the size and shape of each working chamber. The eccentrics, which make the rotor follow the orbiting motion, have two journals and the same eccentricity as the crankshaft. They also drive the valve gear and various accessories.

The idea of having seven combustion chambers was arrived at experimentally. Early attempts were made to build an engine with six chambers but Sarich found it difficult to fix the valve timing. In a four-cycle engine, an even number of chambers gives an uneven or loping firing interval.

The combustion chambers are formed between the rotor, housing and end plates by vanes which do not rotate with the rotor but recede in and out of the outer casing. These vanes are located by slots in the end plates and are held securely as they slide up and down. The ingenious design provides very effective sealing without the use of complex shapes or expensive materials. Seven flats are machined on the rotor's outer periphery to allow the vanes' seals to operate efficiently.

Although the basic principle provides a very compact engine, the main problem was the need to deal with square chambers and the resulting corners. This led to a basically inefficient combustion process. Sealing high-pressure gas was another difficulty. Sarich developed a complex but successful sealing grid designed to bridge the slots in the end-plates which guide the vane legs, forming a continuous end-plate face. The rotor seal runs over this face as the rotor spins. All seals in the engine maintain a fixed angular relationship with their mating surfaces and are lubricated by the crankcase oil. There is no need for oil to be mixed with the incoming fuel charge – a very important point.

The seals are made from conventional chilled cast iron and generally shaped in a simple rectangle. When demonstrating the engine to potential customers, Sarich claimed that the seals were well cooled and did not experience extremes of temperature during each cycle.

The lubrication system serves the bearings and seals and helps to cool the rotor and vanes. The oil supply is through one of the main bearings to the crankshaft from where it is distributed to the other crank bearing and to a rotor spray mounted on the rotating crank. A separate supply feeds the vane cavities. The oil drains to the base of the engine via the outside valve gear enclosure. Scrapers are used to prevent the entry of oil into the combustion chambers.

On most prototypes, conventional spring-and-cam actuated poppet valves were used to open and close the inlet and exhaust ports although

Sarich's rotating disc valve system promised greater efficiency. The first Orbital engine to power a car (a Cortina in 1974) incorporated a system in which disc valves were driven at one-eighth engine speed. However, sealing the discs proved difficult and Sarich realised the concept needed a development program of its own. He decided to persevere with poppet valves because he was more interested at the time in demonstrating the engine's basic principles. The refinements could come later.

The working prototypes had water cooling for the outer housing and end plates. The rotor assembly was oil cooled internally by a controlled bleed from the main eccentric bearing. All major components were made from iron or aluminium with conventional bearing metal used for the eccentric shaft bearings.

The complete Otto cycle of events takes 720 degrees of crankshaft rotation, as in a conventional four-cycle engine. After the fuel is ignited, gas pressure within the working chamber acts on the rotor which in turn reacts on the crankshaft's single eccentricity, creating the output torque.

Sarich experimented with direct fuel-injection from the very start but ran into difficulties and switched to port injection. When further problems were encountered, he fitted a conventional carburettor. He also developed a stratified combustion technique before the first engine was even built.

'The geometry of the engine is such that we can divide the combustion chamber into multiple chambers and have a progression of combustion', he explained.

'In other words, we have progressively leaner zones.'

By 1980, a production version of the E-series Orbital engine with a displacement of 3.5 litres would have weighed an estimated 80 kg, although the prototypes were not so light due to the limited casting and machining processes available to him. The estimated 80 kg was about one-third the weight of a conventional power plant of similar output.

In 1982, Peter Ewing (then an OEC engineer and now a consultant to the company) delivered a paper on the Orbital engine to the Society of Automotive Engineers at an international Congress and Exposition held in Detroit. He said that the engine was deliberately designed to produce maximum torque at a relatively low speed as it was felt that plenty of torque 'low down' was more desirable in everyday motoring than plenty of power at high engine speed. His charts showed that the best output obtained was 90 kW (120 bhp) at 4200 rpm and the torque output was 225 Nm at 2100 rpm. A graph accompanying his paper compared the brake specific fuel consumption of the Orbital (at 1500 rpm) with four other units. The Orbital was 10 per cent better than the 1980 Buick V6. He said that the hydrocarbon exhaust emissions from the engine were higher than desired but that this important area had only recently received attention. However, NOx emissions were low, illustrating one of the engine's advantages.

The basic dimensions for that 1982 unit were given as follows:

number of chambers – 7
stroke – 50.8 mm
chamber width – 110.1 mm
piston (rotor) radius – 128.0 mm
piston (rotor) flats to centre – 122.8 mm
housing radius – 154.2 mm
housing flats to centre – 149.0 mm
vane thickness – 20.0 mm
displacement volume – 3.44 litres

Mr Ewing said the engine had done well in performance trials, logging over 450 hours of vigorous testing without refurbishing. He added that the most critical seals and bearings tended to 'run in' for approximately ten hours of initial operation, after which no perceptible increase in wear was apparent.

Mr Ewing told the Congress that the early development time was mainly confined to producing a durable displacer (i.e. gas pump) because many basic problems had to be overcome first. During early testing, the main bearings continually failed, although the crankshaft was rigid and the bearings carefully line-bored. The failures were caused by crank deflection and Ralph Sarich found an easy solution by introducing flexibility into the main bearing housing support. The trouble was eliminated.

Problems with the eccentrics were also encountered. Several forces which act on the rotor create torque which must be resisted by the eccentrics. Premature failure of the eccentric bearings occurred in all early engines, even though their design adequately catered for the theoretical loading. Tests showed that differences in thermal expansion of the oilcooled rotor and watercooled end-plates caused the failure. Once again, Sarich produced the answer. He fixed the problem by mechanically and thermally isolating the section of the rotor receiving the eccentrics, the new component being called a stabiliser plate.

Until 1980, most new and reconditioned engines suffered from an abrasive wear problem on the major seals and other bearing surfaces which seriously degraded the performance. A detailed investigation isolated the cause as an abrasive material embedded in the metal during the grinding and hand-honing operations. After the machining techniques had been modified, the engine's durability took what Sarich later called a 'quantum jump'.

Only one car was tested with an Orbital engine – a Ford Cortina supplied by Ford Australia in 1974. Sarich also fitted an engine to a Renault 17 provided by Renault Australia but no serious road testing was done on it. Sarich says the Cortina went well but the fuel economy was poor because the engine was not tuned for the installation and the gear ratios were less than ideal.

'From that point on, we became more scientific. We employed a couple of graduate engineers to do basic research into airflow and direct injection. We subsequently developed much improved fuel economy.'

One of the engineers – Peter Ewing – concentrated on basic issues and the second, Kim Schlunke, was told to tackle the breathing problem and see where the restrictions lay. He did so by rigging up a huge Nissan truck Roots supercharger and creating an enormous racket as maximum airflow was forced through the engine.

The research unveiled some interesting facts which showed that the combustion process was unique. Due to the rotor's motion relative to the housing, a fundamental characteristic was discovered. At crankshaft angles prior to and after top-dead-centre (TDC), a large portion of the chamber's charge was transferred from the leading to the trailing edge of the chamber – that is, opposite to the crankshaft's direction of rotation. Although most unusual, this could be made to work to great advantage. Tests also showed that the rotor-to-housing clearance and pocket geometry could be controlled to create a natural division of the chamber near TDC.

In other words, varying the clearance between the rotor and housing at different positions in the chamber tailored the time and intensity of the gas motion to suit the type of combustion required. The transfer of gas within the chamber could be likened to swirl (circular movement of the incoming air/fuel mixture) and squish (motion induced by closing the clearance between the piston and cylinder head) in a reciprocating engine. However, unlike the induction swirl which sometimes occurs in conventional engines, the Orbital's volumetric efficiency and hence output were not compromised by the gas motion.

This attribute meant that a stratified combustion process – that is, one in which a small volume of relatively rich fuel mixture ignites a much larger volume of lean mixture – would work exceptionally well. Tests had already shown that the highly turbulent gas motion allowed lean fuel mixtures in combination with relatively high compression ratios (around 9.3:1). Obviously, direct fuel-injection (forcing fuel into the working chamber) would provide even more benefits.

It is accepted that the theoretically correct air/fuel ratio in an engine, to provide complete combustion of the fuel, is 14.7 parts of air for every one part of petrol, measured by mass. This is known as the stoichiometric ratio. In recent years, ratios as high as 25:1 have been successfully achieved under some conditions and a few computer-controlled engines now run with a ratio somewhat leaner than 20:1. Most, however, go no leaner than 18:1. Stratified combustion promises to make even leaner ratios practical, with resulting benefits for fuel consumption and emission levels.

During the late 1970s, to take advantage of this promise, Sarich recommenced his experiments with fuel-injection. Initially, he used a pump system originally designed for a diesel engine and almost immediately achieved a big reduction in fuel consumption. This opened up an entirely new opportunity. As he sought to further improve the injection system, Sarich refined the two-stream fuel-injection system which today promises to revolutionise automotive practice.

The refinement also led to work ceasing on the Orbital engine concept just at a time when it was achieving a significant advantage over the conventional designs.

Both Ralph Sarich and Kim Schlunke believe that the Orbital engine had the potential to be successfully mass-produced. Sarich says that when they worked on a program for GM in 1982, the Detroit company wanted him to produce a 10 per cent fuel economy advantage over the Buick V6 engine – and this was done. He says that later they got closer to 15 per cent, while maintaining a comparable emission level.

Sarich says that he switched horses when he realised that a three-cylinder, two-cycle engine with the OCP design would produce the lightweight, fuel-efficient unit that he had set out to design in 1970. And to match it with the Orbital design he would need major improvements which were not in sight.

Schlunke agrees but adds:

'The Orbital engine, in the form that was finally developed, had a fuel economy advantage over the equivalent piston engine of its day. It also had a big size and weight advantage.

'However, the sealing grid was complex and difficult to manufacture and I doubt if the engine would have had a cost advantage in volume production.

'If we were to present the 1982 version of the Orbital engine to a manufacturing firm today, it would be difficult for the firm to estimate whether or not it was a viable unit.

'We feel we have all the bases loaded in our favour with the current three-cylinder, two-cycle reciprocating unit because of its low cost and simplicity.

'Yet, if I had to point to the biggest single point in OEC's favour, I would say it is the weight advantage.'

Right: The seven chamber disc valve version of the Orbital engine was completed around 1972. It was promising but unreliable.

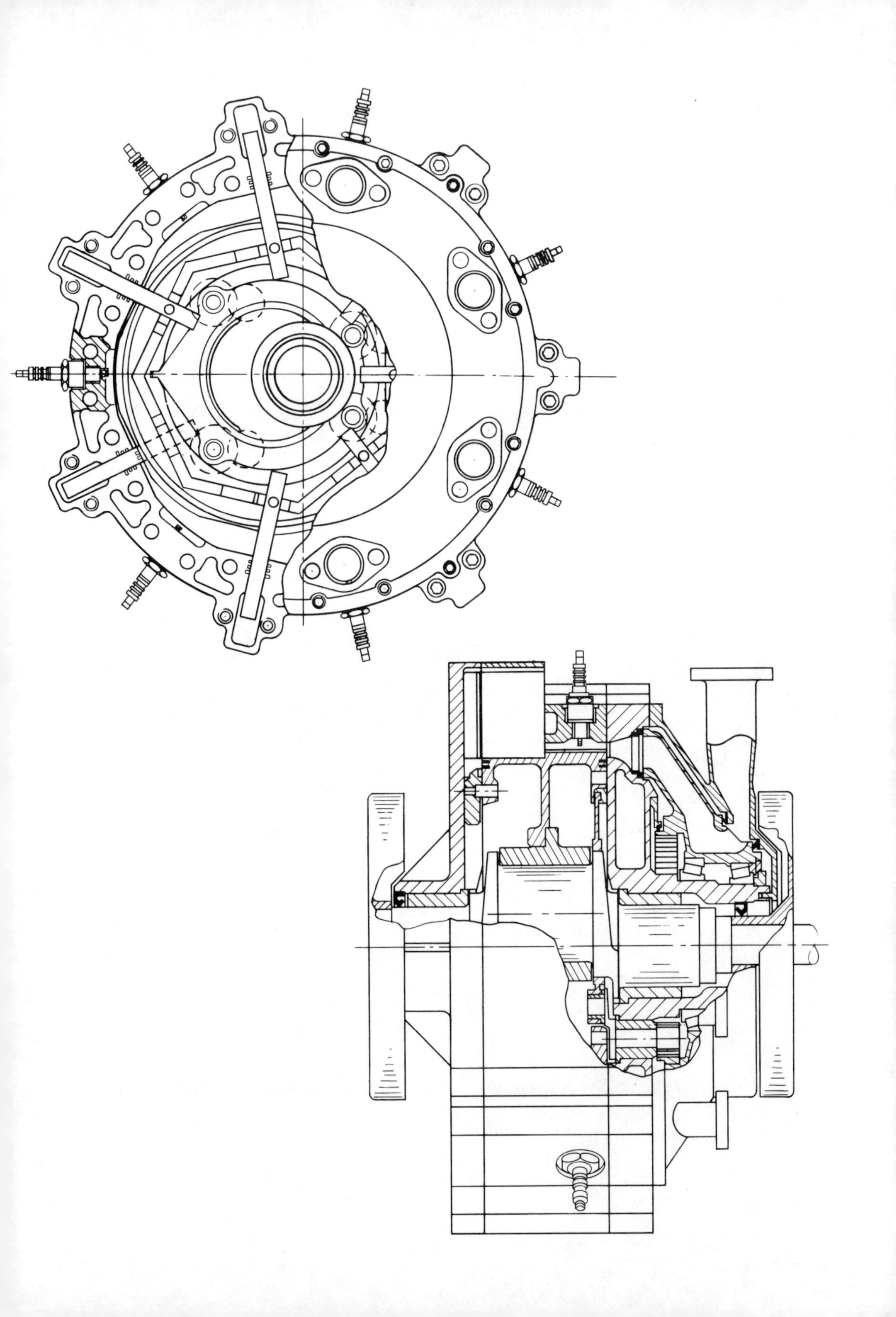

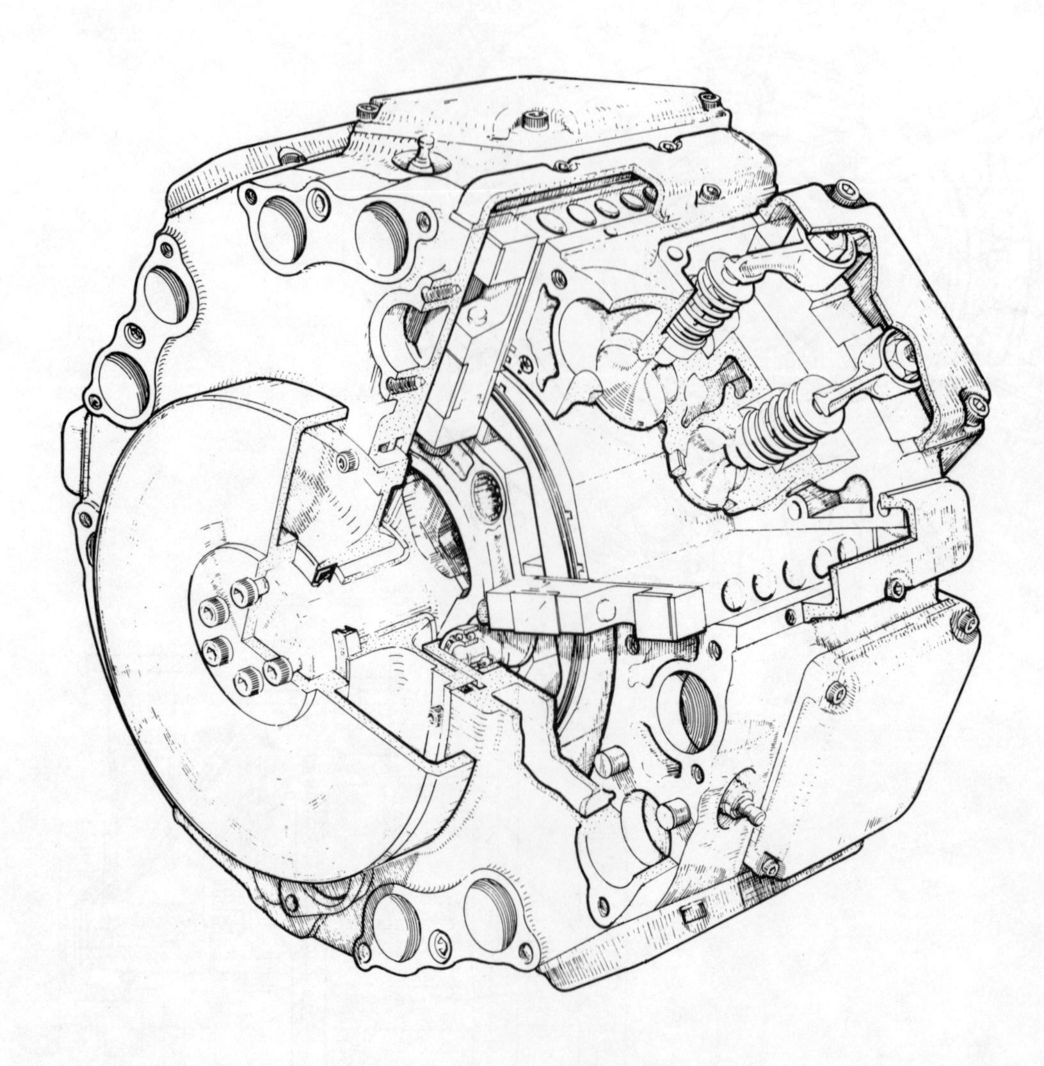

Above and Right: The poppet valve version was more complicated than the disc design but gave less trouble.

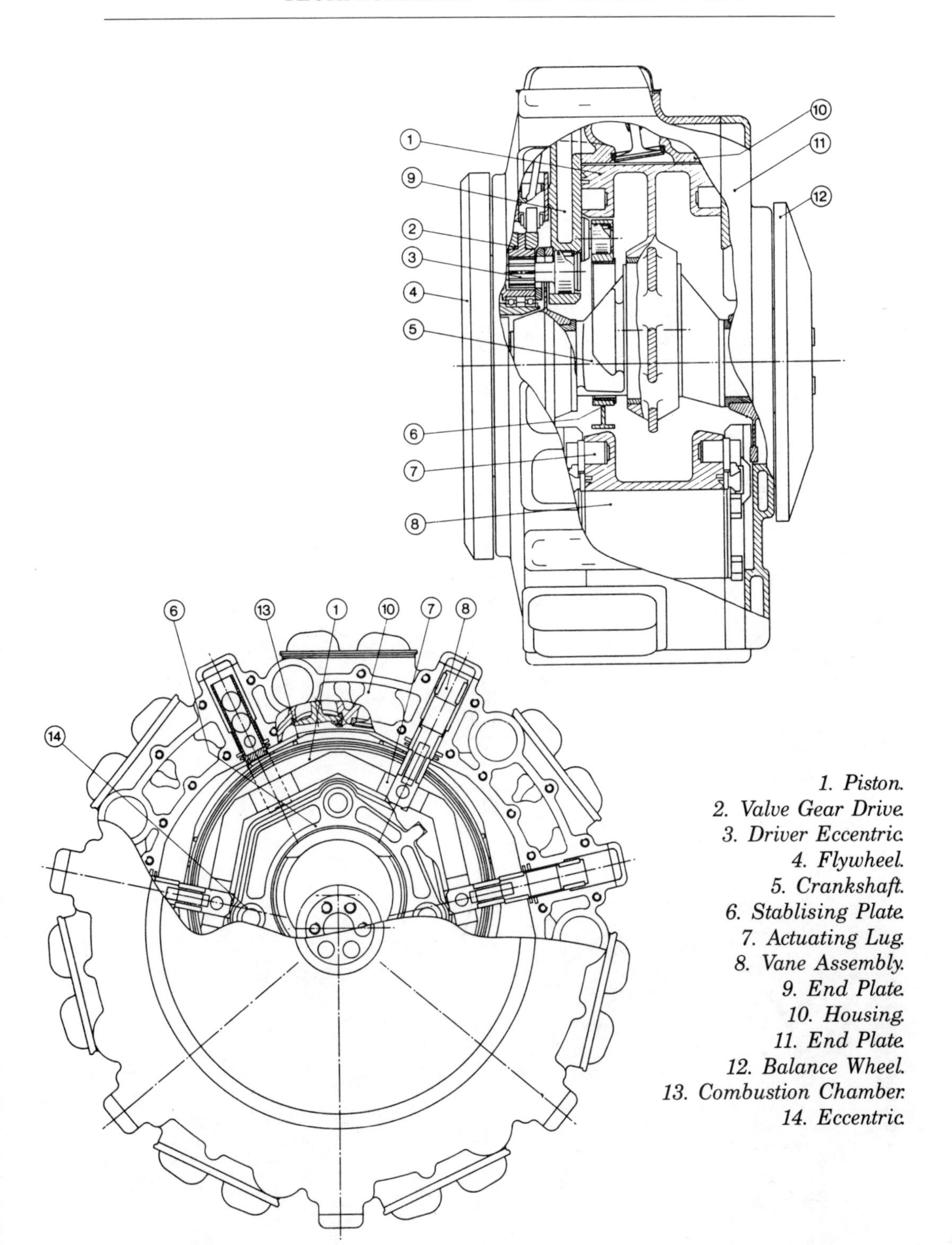
1. Piston.
2. Valve Gear Drive.
3. Driver Eccentric.
4. Flywheel.
5. Crankshaft.
6. Stablising Plate.
7. Actuating Lug.
8. Vane Assembly.
9. End Plate.
10. Housing.
11. End Plate.
12. Balance Wheel.
13. Combustion Chamber.
14. Eccentric.

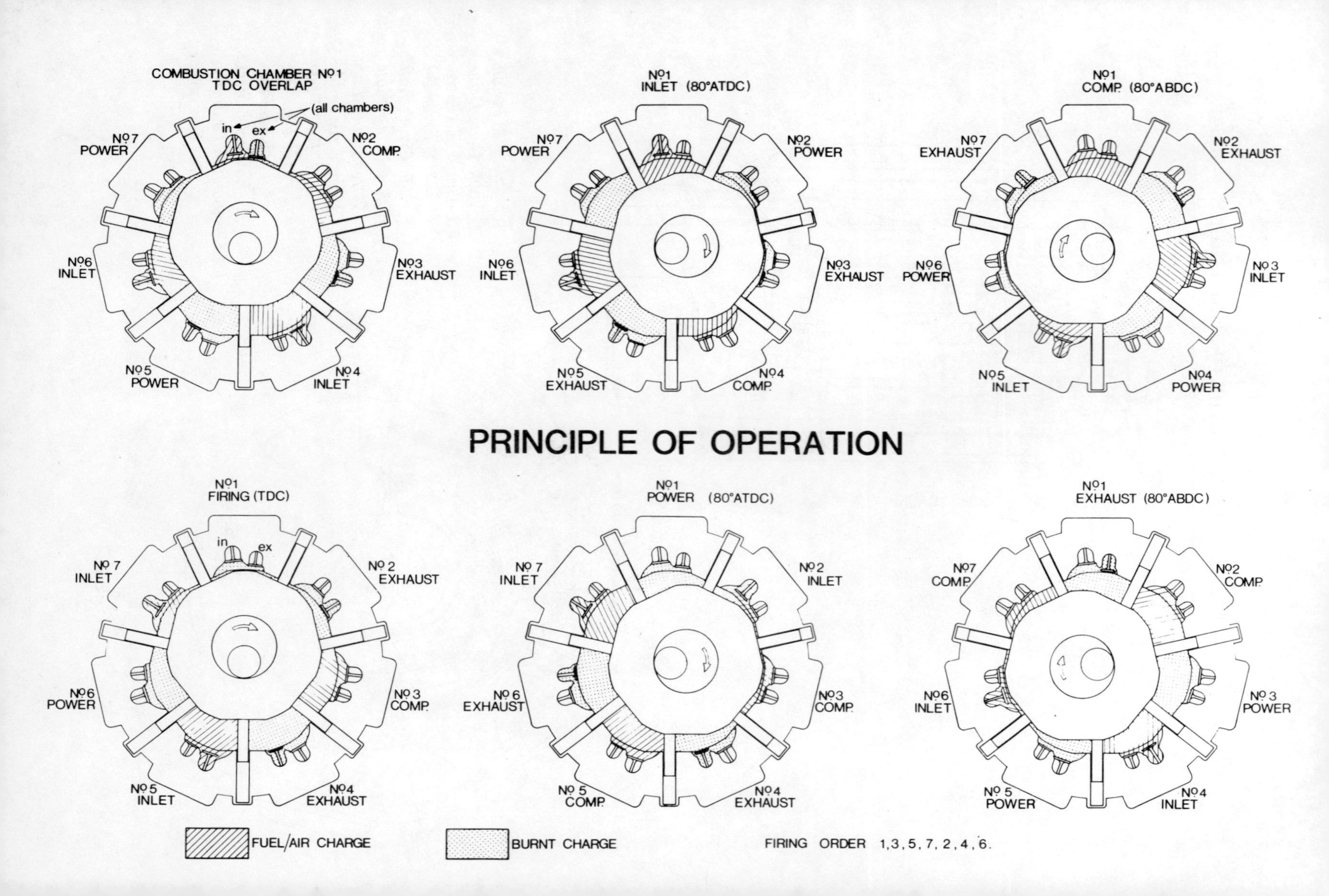

COMBUSTION CHAMBER Nº1
TDC OVERLAP
(all chambers)
in
ex
Nº7 POWER
Nº2 COMP.
Nº6 INLET
Nº3 EXHAUST
Nº5 POWER
Nº4 INLET
Nº1 INLET (80°ATDC)
Nº7 POWER
Nº2 POWER
Nº6 INLET
Nº3 EXHAUST
Nº5 EXHAUST
Nº4 COMP.
Nº1 COMP. (80°ABDC)
Nº7 EXHAUST
Nº2 EXHAUST
Nº6 POWER
Nº3 INLET
Nº5 INLET
Nº4 POWER
PRINCIPLE OF OPERATION
Nº1 FIRING (TDC)
in
ex
Nº7 INLET
Nº2 EXHAUST
Nº6 POWER
Nº3 COMP.
Nº5 INLET
Nº4 EXHAUST
Nº1 POWER (80°ATDC)
Nº7 INLET
Nº2 INLET
Nº6 EXHAUST
Nº3 COMP.
Nº5 COMP.
Nº4 EXHAUST
Nº1 EXHAUST (80°ABDC)
Nº7 COMP.
Nº2 COMP.
Nº6 INLET
Nº3 POWER
Nº5 POWER
Nº4 INLET
FUEL/AIR CHARGE
BURNT CHARGE
FIRING ORDER 1, 3, 5, 7, 2, 4, 6.

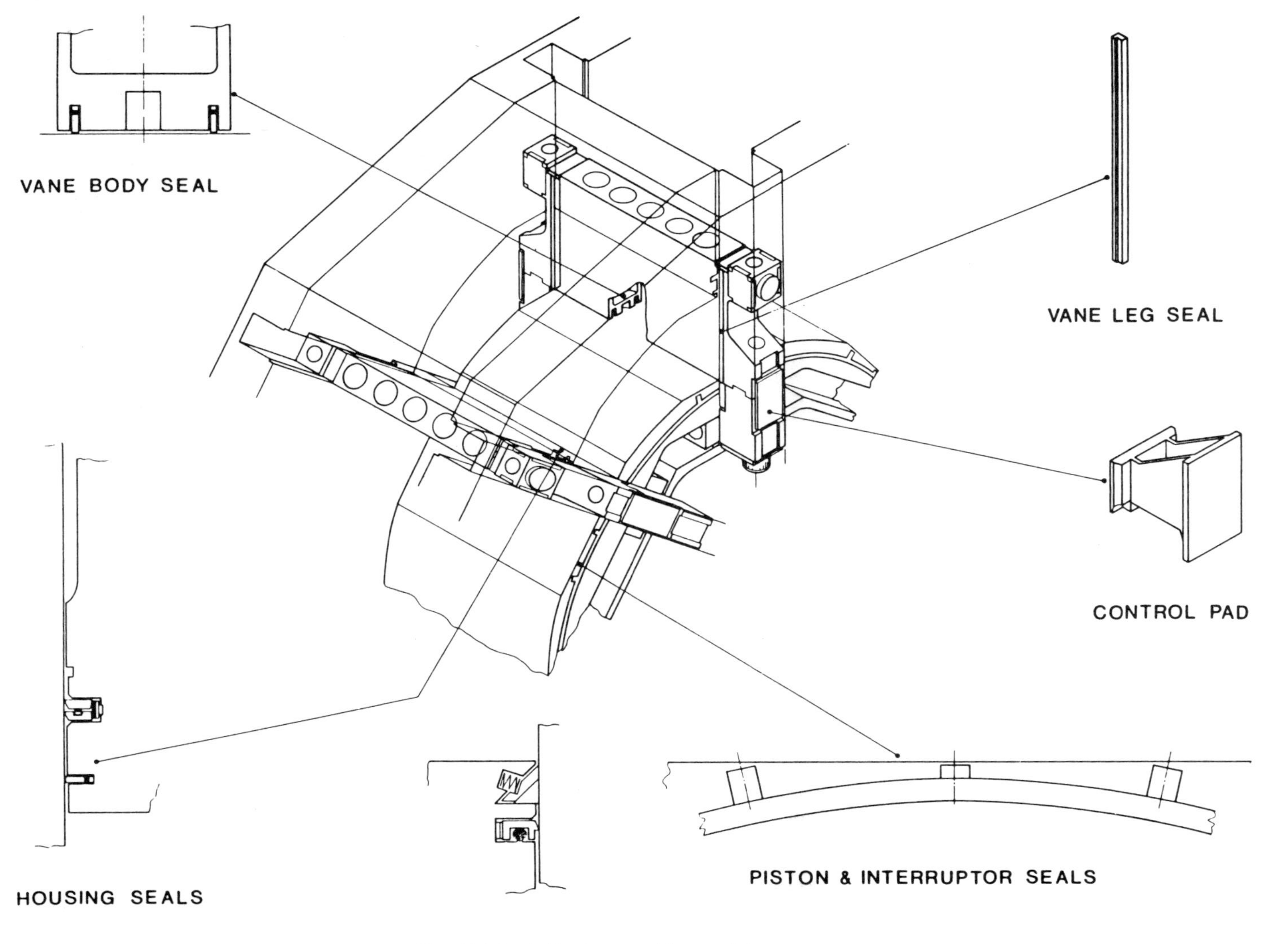

The Orbital Engine sealing grid

3·5L 7 CHAMBER ORBITAL ENGINE
3·5L V8 RECIPROCATING ENGINE

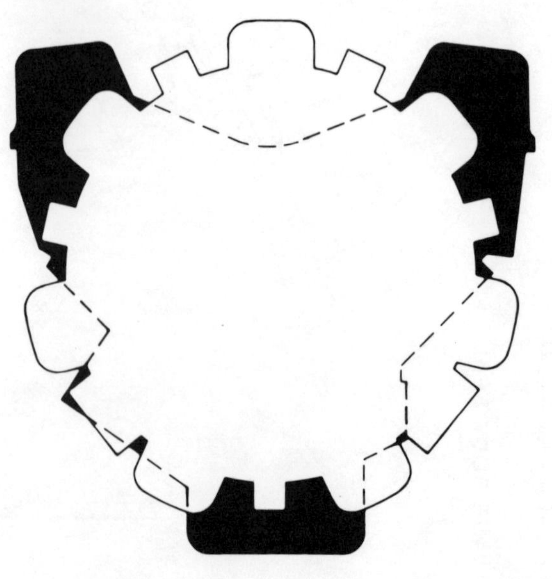

A comparison between the package sizes of the 3.5-litre Orbital engine and a 3.5-litre reciprocating V8.

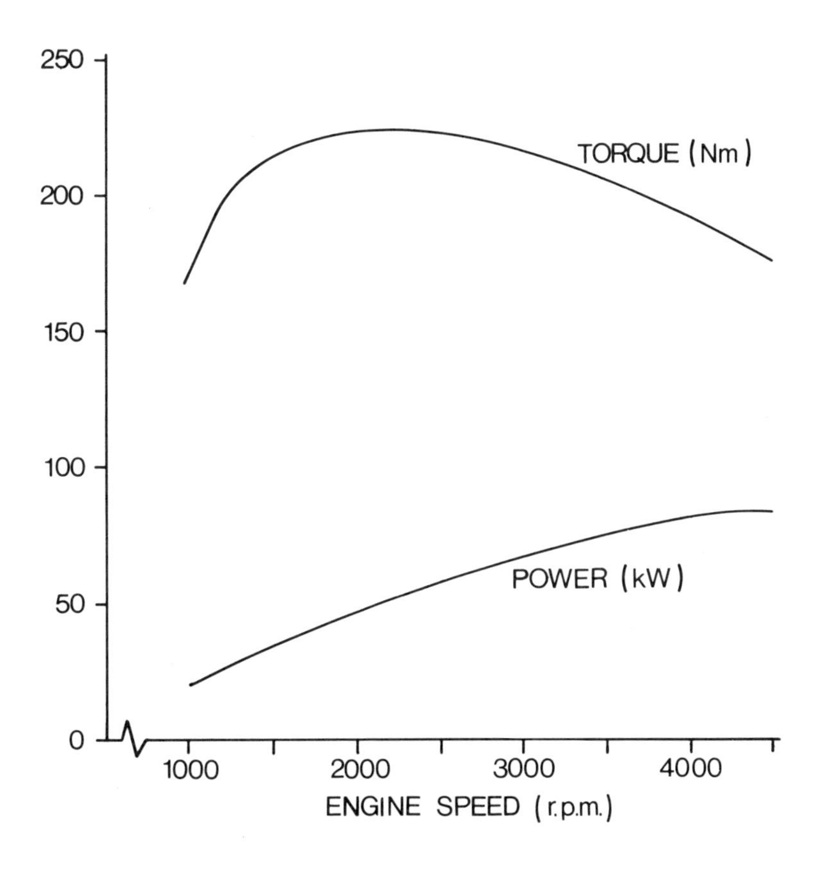

The torque and power characteristics as measured for the D-series Orbital show the excellent torque characteristics at low engine speeds.

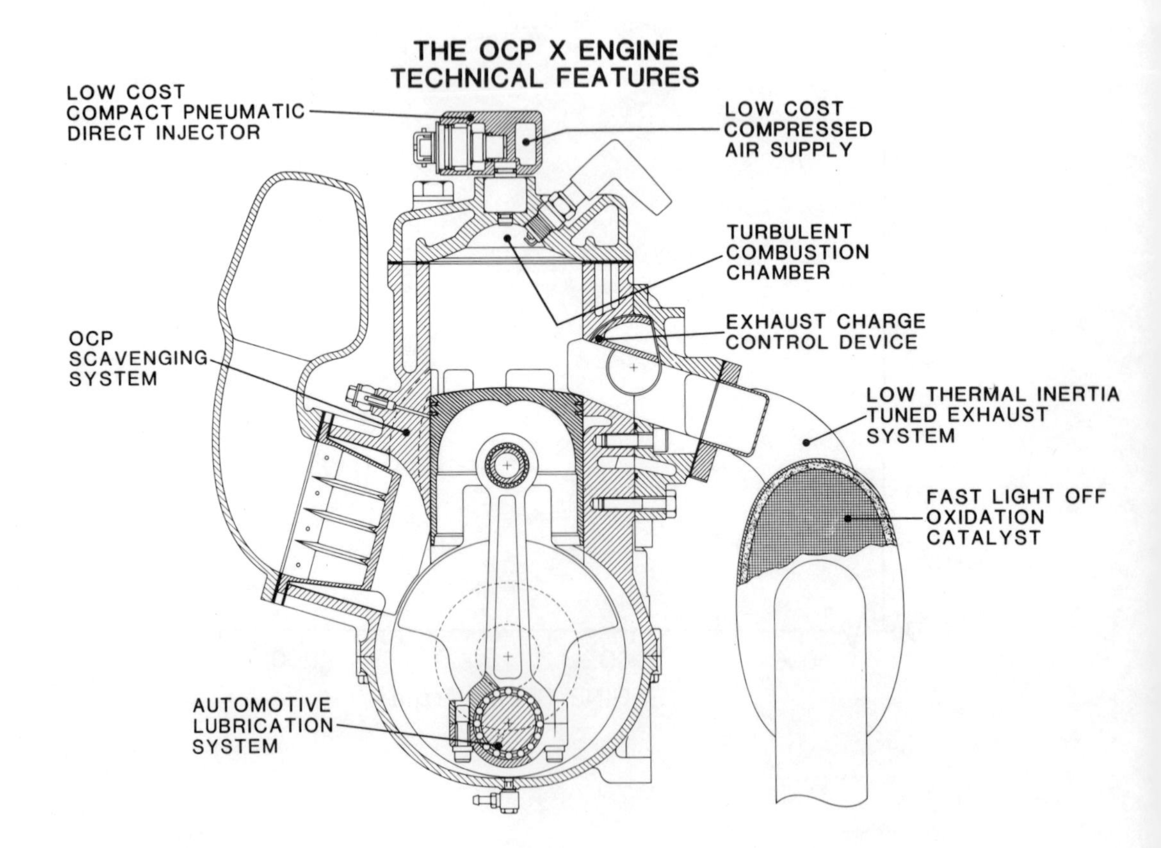
THE OCP X ENGINE
TECHNICAL FEATURES
LOW COST
COMPACT PNEUMATIC
DIRECT INJECTOR
LOW COST
COMPRESSED
AIR SUPPLY
TURBULENT
COMBUSTION
CHAMBER
EXHAUST CHARGE
CONTROL DEVICE
OCP
SCAVENGING
SYSTEM
LOW THERMAL INERTIA
TUNED EXHAUST
SYSTEM
FAST LIGHT OFF
OXIDATION
CATALYST
AUTOMOTIVE
LUBRICATION
SYSTEM

12

Technicalities — the two-cycle OCP engine

Ralph Sarich knew from the day he started to build the first Orbital engine that the key to efficiency lay in the combustion process. It was essential to find a way to make the fuel burn quickly and completely. He was familiar with the work which had been done on stratified combustion and had studied the advantages of mechanical and electronic fuel-injection systems. He began to ponder the merits of both forms of technology and, with typical disregard for conventional wisdom, explored all possible routes that the processes could take.

The use of a stratified combustion process — which allows a leaner overall fuel mixture to burn — was the obvious way to go. However, others (notably Ford, GM and Texaco) had run into a brick wall in their attempts to produce systems suitable for mass production. Obviously, Sarich had to find an answer to the problems that had scuttled the earlier projects.

Fuel-injection in a petrol car dates to the 1904 US-built Pope. By the mid-1970s, more and more car makers were switching from carburettors to fuel-injection to reduce exhaust emissions and enhance the engine's power. As the name implies, injection is a positive way of getting fuel into the cylinders, compared with the carburettor route in which the fuel mixture is drawn in by atmospheric pressure. Virtually all fuel-injection systems have relied on mechanical pumps which force the fuel through injectors into the inlet manifold or directly into the combustion chamber.

Sarich was keen to use injection because the process is more controllable than a carburettor and comes closer to the ideal. The main advantage of fuel-injection is that very precise amounts of fuel can be delivered, and using computer controls or mechanical devices, the fuel can be metered to suit varying operating conditions. This means injecting different quantities of fuel when idling, accelerating, steady speed cruising or travelling downhill with the foot off the accelerator pedal (i.e. travelling on the overrun).

There are basically two types of fuel-injection. The easier and less expensive conventional approach is to inject the fuel into the inlet manifold and allow it to be drawn into the cylinders when the inlet valves open. Potentially more efficient (but more difficult) is the alternative 'direct injection' route in which fuel is forced directly into the pressurised combustion chamber.

Sarich knew that most diesel engines have direct fuel-injection and that Rudolf Diesel, the inventor of the diesel engine, had experimented with an injection system which used compressed air blown over a reservoir of fuel. Other engineers have since worked on similar lines but none of these 'pneumatic' systems has reached production.

He was also aware that extremely few gasoline engines have direct fuel-injection because of the extra problems involved. The major drawback is that bulky and expensive pumps are needed because the fuel has to be injected at high pressure in order to achieve atomisation in a chamber where high pressure already exists. He decided that the best avenue of approach was to find a less expensive form of direct injection.

His solution was as brilliant as it is simple. Instead of using a pump to force the fuel into the combustion chamber, he carried the fuel on a stream of compressed air. Unlike Rudolf Diesel, Sarich decided to blow the air through the fuel by creating a small holding chamber with an orifice at the bottom. The compressed air literally shears through the fuel as it goes into the orifice, taking finely atomised fuel with it.

The more he thought about this 'two-stream' approach, the more Sarich liked it.

Apart from the huge reduction in cost and weight, the idea promised several major advantages. Not the least is that the compressed air helps atomise the fuel and creates considerable turbulence within the combustion chamber. In automotive terminology, turbulence is rapid smaller-scale gas motion, often deliberately induced, which ensures violent movement of the air and fuel droplets, causing them to mix evenly and quickly. Turbulence also speeds up the combustion process once the fuel has ignited. The two-stream air system allows the use of much lower fuel pressures than diesel systems and easier interfacing with an electronic management system.

Both the injection system and combustion process developed for the Orbital engine were retained for the reciprocating units and this is why the rather confusing use of the Orbital Combustion Process (OCP) was carried over and used on the new generation of reciprocating engines.

In retrospect, the 'two-stream' injection system idea is almost alarming in its simplicity and engine specialists around the world must be kicking themselves for not thinking of it first. However, like many simple ideas, an immense amount of detail work was required to make it work effectively.

Work on the basic idea started around 1971 and the system was tested, without success, on the first Orbital engine. It was in 1977, when Mike Mckay and Ken Johnsen, working under Ralph's direction, came up with hardware which finely atomised the fuel, that success was in sight. More ideas began to flow and tests showed the two-stream injection dramatically improved the Orbital engine. The second major spin-off emerged when the system was tested on conventional two-cycle and four-cycle reciprocating engines.

It was initially fitted to an Escort Kent engine around 1981. Several Holden Camiras were equipped with direct and indirect injection between early 1982 and 1985. Tests on a two-cycle application with direct injection then demonstrated a fuel economy advantage around 30 per cent and the exhaust emissions comfortably met Australian legislated requirements.

'We were impressed by the extremely low hydrocarbon emissions', says Schlunke.

'During the first few weeks, even before we had the system refined, the hydrocarbons were extraordinarily low. So was the fuel consumption.

'The two-fluid injection was used in the manifold initially for reciprocating engines and in-cylinder for the Orbital engine. Its ability to work in-cylinder is its real strength. Good atomisation is important to manifold injection but it is critical when used in-cylinder.

'When you inject directly into the combustion chamber, you limit the time available to mix the fuel and air stream. In the usual carburettor set-up, the fuel and air mingle during the relatively long time it takes for them to travel from the carburettor to the combustion chamber, so they are usually well mixed by the time compression takes place.

'In a two-cycle engine, you have only the time from when the fuel meets the fresh air in the cylinder to when ignition occurs. The period is so limited that the fuel preparation process is of the utmost importance.'

As developed for the Orbital engine (and later refined for reciprocating units), the OCP design uses a small engine-driven air pump and a specially designed fuel injector. The combustion chamber is shaped to minimise emissions and fuel consumption.

The OCP-designed direct injector delivers extremely small droplets of fuel with a controlled spray geometry directly into the cylinder. The droplet size in a typical diesel fuel-injector is around 40 microns (a micron is one-thousandth of a millimetre). Even in the early days, tests by the Sarich organisation showed that, by using compressed air, the droplet size of the fuel could be reduced to around 10 microns. The size was further reduced as development proceeded. According to Kim Schlunke, typical applications run at pressures around 5 bar, or 500 kPa. The current air pump draws about one per cent of the maximum power output at any given speed.

'An important feature of the tiny droplet size our system produces', says Schlunke, 'is the way it is is maintained throughout the spray plume. Analysis with our laser particle sizer shows that the droplet size distribution remains much the same throughout the spray, indicating there is very little coalescing of the droplets towards the end of the spray plume. The size distribution and spray geometry help control hydrocarbons and NOx [oxides of nitrogen] emissions.'

In conventional engines burning unleaded fuel, expensive, three-way catalytic converters are required to convert the major pollutants to water, carbon dioxide, oxygen and nitrogen. With the Orbital Combustion Process, as applied to a two-cycle engine, the OEC engineers found that they had fewer emissions to treat and could handle them with a low-cost oxidation catalyst. The OCP catalyst uses platinum and paladium, which saves money because rhodium, an expensive metal, is not required as in a three-way catalyst.

Until this new OCP system was developed, the two-cycle engine could not be realistically considered for modern cars because of the excessive emission of pollutants. The hydrocarbon and carbon monoxide emissions were typically ten times the required levels and the fuel consumption twice that of a conventional four-cycle unit. In May 1987, an OCP two-cycle engine, installed in a vehicle and calibrated for compliance, was evaluated over the US EPA CVS cold/hot test using the LA4 driving cycle. It provided a substantial engineering margin under the current US limits, even though the NOx was not treated catalytically. The unit was only fitted with an oxidation catalyst which used platinum and paladium.

Five years earlier, OEC had established that healthy commercial prospects existed for the combustion concept and orders were received for prototype engines and fuel systems from major car and marine engine manufacturers in the USA, Europe and Japan. By then, the initial work on the four-cycle reciprocating engine was being eclipsed by remarkable successes achieved with two-cycle designs.

OEC engineers were thrilled by the way it eliminated the basic drawbacks of the two-cycle design without affecting the advantages. From then on, they homed in on this area. If they could match the virtues of the four-cycle design with a unit offering lower weight, reduced manufacturing cost and compact size, the design would be almost irresistible for the mass market. Unlike the Orbital engine, a reciprocating unit appeals to major manufacturers because it employs conventional components which they are already geared to manufacture. These include the cylinder block, pistons and crankshaft.

Early in the program, OEC engineers realised that although the process is suitable for an engine of any size, the use of three cylinders would provide cost, packaging, weight and size advantages over a 'four'. The configuration also allows optimum exhaust manifold configuration.

The exhaust system of a three-cylinder engine does not require a complex system of tuned lengths to achieve a good, broad torque range. The typical port timing of a three-cylinder, two-cycle unit is that the exhaust blows down 90 degrees after top-dead-centre and the process at low speed occurs over a period of 30 to 45 degrees. That is extremely desirable because 120 degrees disposed from that blow-down process the engine is closing up the exhaust port in an adjacent cylinder. This means that the blown gas from one cylinder helps to charge the adjacent cylinder.

In other words, the high exhaust pressure from one cylinder can force gas along a duct into the cylinder which is 120 degrees displaced from it. That happens to be the exact time when it is needed. The transfer port has just closed and the mixture is being forced out of the chamber because the piston is just rising. You don't want the mixture to be lost and, fortunately, a high-pressure pulse arrives at the port just in time and forces the mixture back in.

'The beautiful part of the process is that the effect occurs right across the speed range – and this makes the three-cylinder set-up an optimum configuration', says Schlunke.

'The exhaust plumbing on a three-cylinder engine is easy to make and can be fabricated from low-cost stampings. It also has a very low thermal inertia which is important because it helps 'light off' the catalyst [i.e. rapidly bring the catalyst to its operating temperature after a cold start].

'We have added a variable valve-timing device which allows the engine to automatically make subtle adjustments to the timing. It is driven by a small DC motor controlled by the engine's electronic control unit and acts as a variable height exhaust port.

'Some motorcycle makers have used a similar mechanism in their engines as a power valve but ours is for emission control reasons although it contributes to the very flat torque curve. It provides a novel and very simple means of dynamically tuning the exhaust during engine operation.

'It helps provide one of the key features of the OCP system – our ability to retain the exhaust gas. This is done electronically, coordinating the inlet valve, throttle valve and exhaust valve, and is the reason we don't have the characteristic corn-popping misfire noise when you back off the accelerator – as in most two-cycle designs.

'Our engine does not misfire because the charge is stratified and continues to burn even when a lot of exhaust gas is in the cylinder.

'The desirable by-product of keeping some exhaust gas in the cylinders is that we retain the heat of the gas and this raises the local lean flammability limit of the charge.

'Now a stratified engine goes from a rich mixture at the spark plug to a leaner condition at the outer limits of the cylinder. That's what is different about a stratified charge engine compared with a pre-mix or homogenous mixture design.

'The rotten feature of most experimental stratified engines is that the flame burns through the variable air/fuel ratio and reaches a figure of about 100 to 1 – and this is called the lean flammability limit – where it can no longer support a flame and becomes extinguished.

'There is still fuel in the area through which the flame has not passed and it constitutes a major part of the hydrocarbon exhaust emissions.

'If you could raise the temperature of that charge, the flame would burn further because, as the charge temperature goes up, the lean flammability limit can be extended. If the flame can travel further, it would consume most of the fuel which might otherwise have gone out with the exhaust.

'This is where our variable exhaust port fits into the scheme of things. We extend the lean flammability limit by using very hot exhaust gas to raise the local charge temperature in the cylinder. A bonus to the system is that hydrocarbons which might have gone out in the exhaust system get a second chance to burn.

'This is a major breakthrough ... our ability to control hydrocarbon emissions in a stratified engine. The concept works in both two-cycle and four-cycle units.

'Our second major technical achievement is the ability to control NOx. A lot of research into stratified engines has not reached the point where NOx becomes a problem. NOx occurs because, if a charge is stratified, the first part of the charge to burn can have exactly the wrong sort of stoichiometry [air/fuel ratio] to keep down NOx.

'People have known for years that if you run a normal carburettor pre-mix engine with an 18 to 1 air/fuel ratio it produces copious quantities of NOx.

'In a stratified engine you can reduce that condition locally at the spark plug. In other words, the overall air/fuel ratio might be 30 to 1 – which is very, very lean – but if the condition around the plug is, say, 18 to 1 and that gas happens to be at the highest temperature for the longest time, you are certain to generate lots of NOx. You are emulating the bad position of the conventional engine.

'The special thing about our engine is that we can control the stratification gradient [the air/fuel ratio of the mixture] as the flame approaches it. Every engine has a characteristic burning rate which can be described as a curve which plots the mass that burns against the crank angle. Each segment of mixture has a different air/fuel ratio and we have learned how to control each individual ratio electronically.

'We can also ensure that the charge at the spark plug is always rich so we never produce large quantities of NOx.

'In the OCP engine, the average mixture for cruising is between 20 and 25 to 1 but for light load running it can be as lean as 60 to 1.

'Mostly, the mixture is around 20 to 25 to 1, but for very cold starts, it could be 50 to 1 and for slightly warmer starts 40 to 1.'

OEC has been working on the two-cycle unit for five years, a very short time by automotive standards. The original two-cycle prototype was based on a three-cylinder engine block taken from a Suzuki outboard motor. Even the earliest tests thrilled the Sarich team. In addition to the reduced weight, package size and superior fuel economy, the 1.2-litre capacity unit yielded slightly more power than a conventional 1.6-litre, four-cycle unit along with lower exhaust emissions.

The output has since been increased to 78 kW (105 bhp) gross at 5500 rpm, with a maximum torque of 138 Nm at 4000 rpm. This makes the specific output around 40 per cent higher than that of a typical modern four-cycle engine of comparable capacity.

Very early in the program, the OEC engineers concluded that the technology is suitable for any petrol engine from 50 mL (50 ccs) upwards, whether operating on a four-cycle or two-cycle principle although the advantages are far more pronounced when it is applied to a two-cycle design, preferably with three cylinders or multiples thereof.

There are several other interesting developments on the way.

Work is being done to get rid of the engine-driven compressor by using captive pressure straight from the combustion chamber to carry fuel into the cylinders. OEC engineers initiated the idea but have it on the back-burner at present because of more urgent development priorities. However, a customer is currently testing an OCP engine with a captive pressure system known as CPES. This ingenious idea has great promise and OEC engineers have assigned it a catchy in-house name. They call it a Clayton compressor, a play on the Clayton advertisement for the 'drink you have when you are not having a drink'.

Likely to reach production ahead of CPES is a supercharged OCP engine.

To understand why a supercharger is used and not a turbocharger, apart from cost considerations, it is best to recall that the conventional two-cycle engine has internal scavenging in which air is drawn into the crankcase and pumped by the lower side of the piston from the crankcase into the upper cylinder. To have forced induction, it is necessary to turn to an externally scavenged design. The idea is not novel and many diesel engines have external scavenging with some kind of blower to pump the air through the engine. As Kim Schlunke explains, the normal turbocharger set-up is not suitable.

'You can't use a turbocharger alone to perform this job. To force the air in, the pressure has to be lower on the downstream side so that air can go into one side of the engine and out the other.

'Generally speaking, you would not be able to operate the engine using a turbocharger alone because the pressure coming into the turbocharger is normally higher than the pressure coming out of its compressor and this makes it impossible to drive the air across the two-cycle engine. However, it is certainly possible to achieve a positive scavenge gradient under some conditions.

'It would be hard to start the turbo spinning unless you have hot high-pressure gas going into it, although there are devices around which would let you start the turbo spinning. An alternative is to use a supercharger with the turbocharger. The GM two-stroke diesel, for example, uses a Roots-type blower downstream and a turbocharger upstream. Of course this adds to the price. As a rough industry number, a turbocharger adds $200 to the engine manufacturing cost but we can put a belt-driven centrifugal compressor there for considerably less – and still get the same sort of performance as we would with a turbo.'

Another interesting development is a 900 mL (0.9 litres) variant of the standard engine. The current 1.2-litre unit develops 78 kW (105 bhp) at 5500 rpm but the 900 cc unit is expected to develop nearly as much power because the operating speed could be pushed to 9000 rpm.

This would be a great engine in markets where the under 1-litre category is a distinctive class of its own. Several high-tech conventional engines are coming to the under 1-litre class, most of which develop a high output at high speed.

'Our engine does not have the usual valve train limitations and would be able to rev to high speeds without sacrificing low-speed performance', says Kim Schlunke. 'We'll be in there beating the best fuel economy engines in the world.'

COMPARATIVE TEST

An early 1.2-litre, three-cylinder, crankcase-scavenged, two-cycle OEC engine, burning unleaded fuel and fitted with a oxidation-only catalyst, was installed in a Holden Camira J-car. The overall engine package was 50 kg lighter than GMH's standard 1.6-litre, four-cylinder, four-cycle engine and essentially half the packaging size.

The chart compares a fully equipped 1.2-litre OEC engine with a standard 1.6-litre Camira. Figures, where possible, are included for a 1986 JD Camira 1.8-litre engine with throttle body injection.

FUEL CONSUMPTION (litres per 100 kilometres)

	city cycle	country cycle
1.2-litre OCP engine	6.8	5.9
1.6-litre carb GMH engine	9.5	6.5
1.8-litre TBI GMH engine	9.5	6.8

The 27 per cent improvement in fuel consumption was realised over the standard AS 2077 city driving cycle while meeting ADR 27A emission levels.

EXHAUST EMISSIONS (grams per kilometre)

	ADR 27A			ADR 37		
	HC	NOx	CO	HC	NOx	CO
1.2-litre OCP engine	0.95	0.57	0.43	0.90	0.67	0.30
1.6-litre carb GMH engine	1.30	1.20	15.0	1.30	1.35	13.0
1.8-litre TBI GMH engine	0.68	1.72	3.90	0.60	1.70	3.30
legislated level	2.10	1.90	24.2	0.93	1.93	9.30

ORBITAL COMBUSTION PROCESS ENGINE

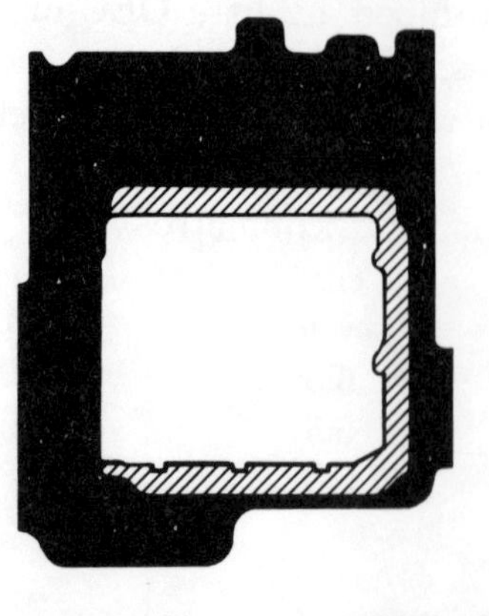

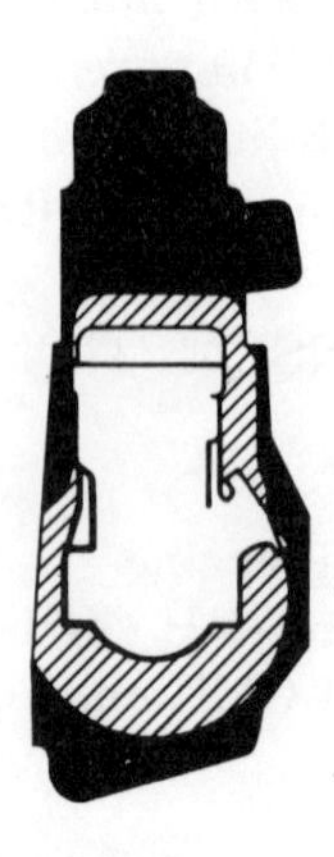

1.2 LITRE OCP 'X' ENGINE
1.2 LITRE OCP 'C' ENGINE
1.6 LITRE GM WORLD CAR

Above and Following Pages: Comparative information and dynamometer test results for an OCP engine were released by the Orbital Engine Company in 1987. Numerous improvements have since been made. Two 1.2-litre OCP units (C-series and X-series) are compared with a 1.6-litre GM 'four'. The OCP has a big advantage in fuel economy at the light loads and speeds which are important to the driving cycle. BSFC stands for brake specific fuel consumption which is plotted against the load. ITWC indicates GM 'Family II' engine.
The OCP engine was evaluated over an EPA CVS C/H test using the LA4 driving cycle with an oxidation catalyst. The fuel consumption was 10 per cent less than the 'best in class'. The pumping work diagram (page 215) illustrates the basic advantage of a two-cycle engine compared with a four-cycle design, hence the good fuel economy.

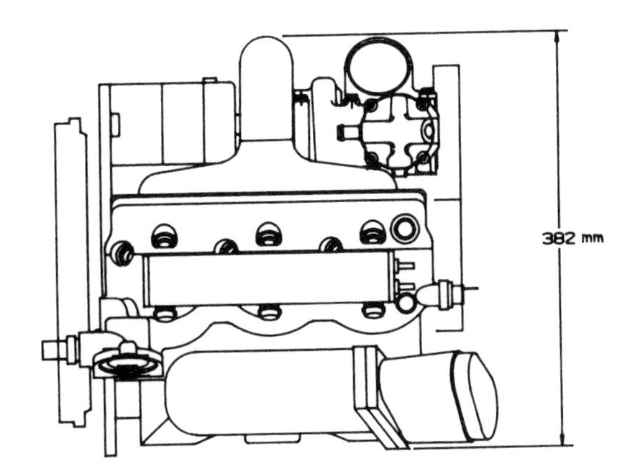

SCHEMATIC OF XIHB ENGINE

- 1.197 LITRE DISPLACEMENT
- 84mm X 72mm BORE & STROKE
- POWER 65 kW @ 5500 RPM
- TORQUE 140 Nm @ 3000 RPM
- ESTIMATED WEIGHT 51 kg AS SHOWN
- ALUMINIUM HEAD,BLOCK & CRANKCASE
- ALUMINIUM OR COATED BORE
- 100mm BORE CENTERS
- ONE PIECE CRANKSHAFT
- FRICTIONLESS BEARINGS THROUGHOUT
- PARALLEL COOLING
- SCAVENGING AXIS NORMAL TO LONGITUDINAL AXIS

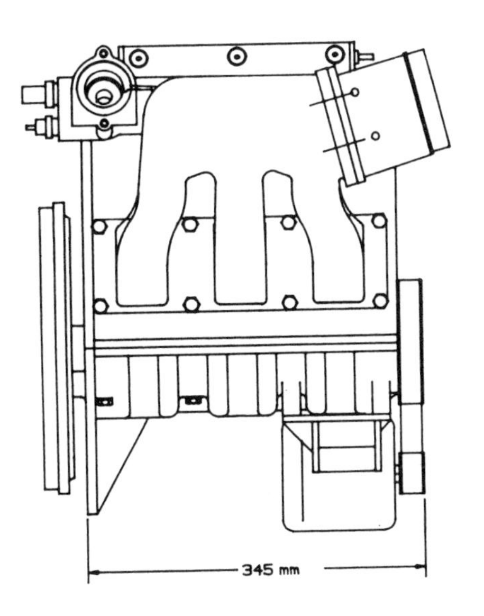

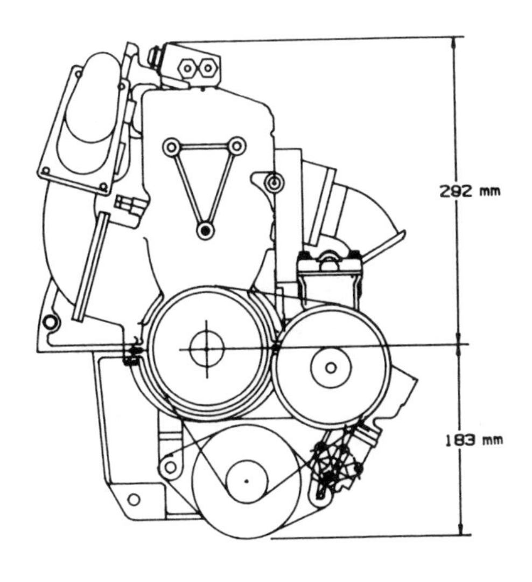

1987 Sports Coupe with X engine in equivalent position.

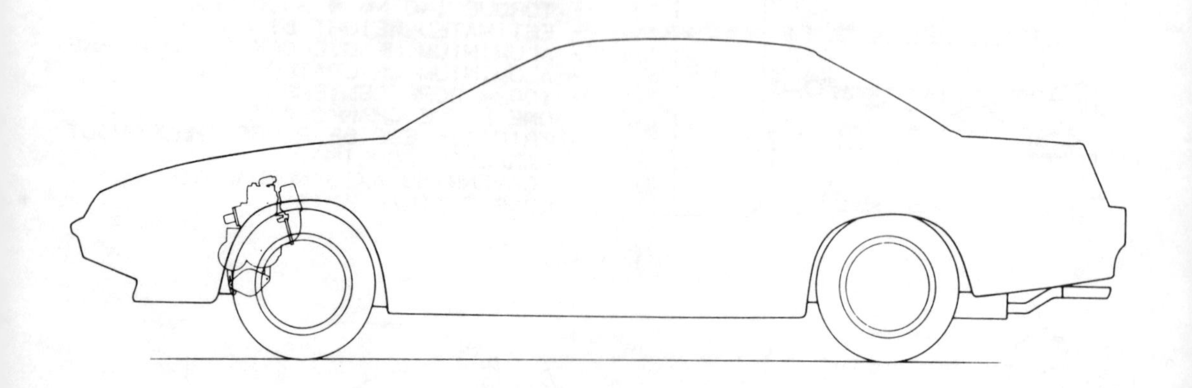

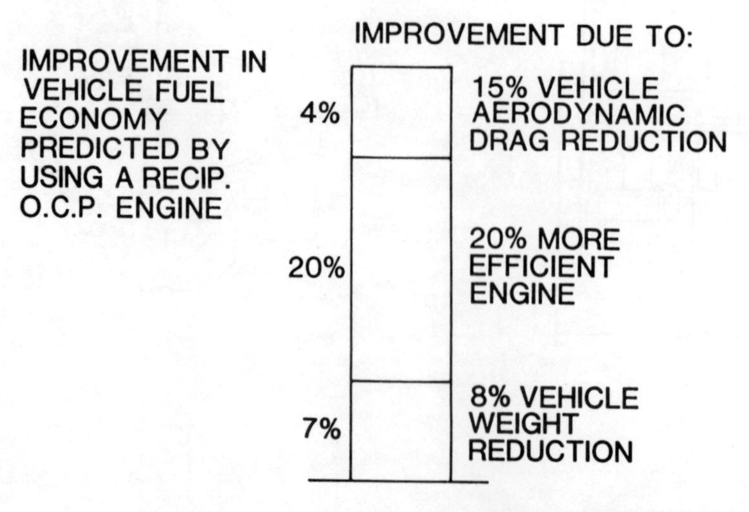

VEHICLE FUEL CONSUMPTION
EPA URBAN CYCLE

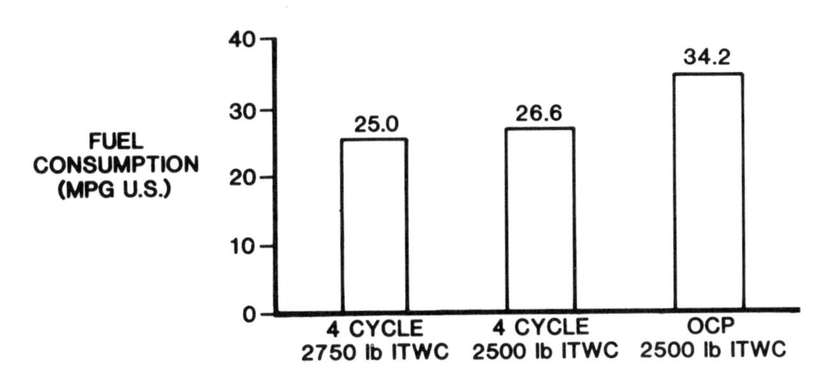

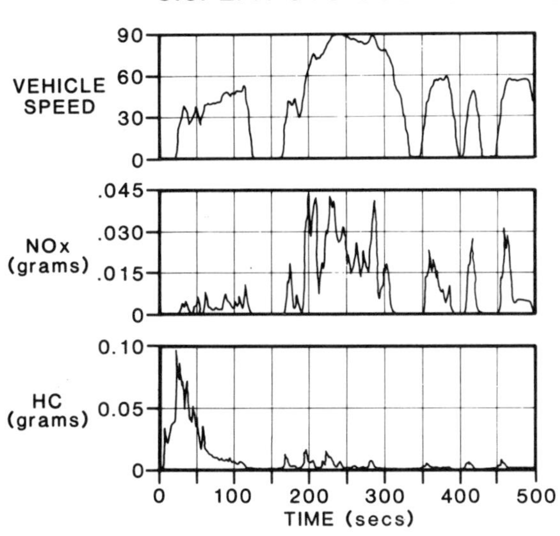

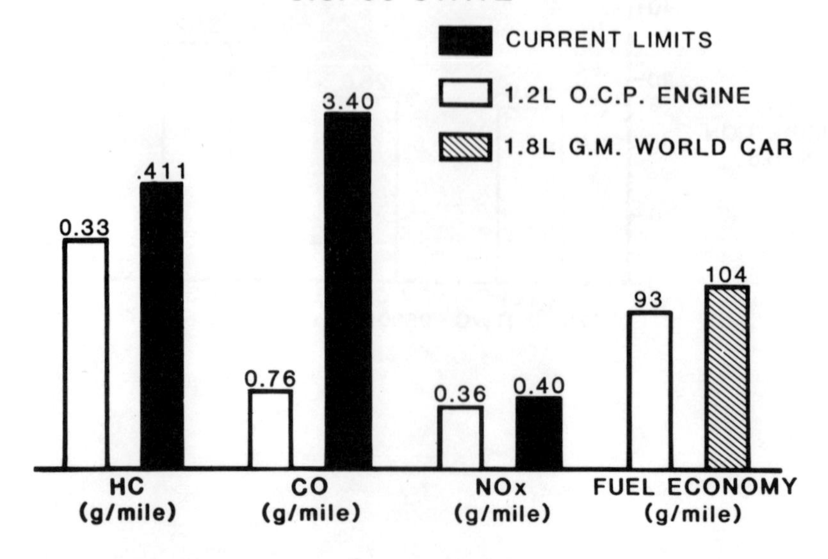
E.P.A. URBAN DYNAMOMETER SCHEDULE
FOR 2625 lb ITWC - PRESENT STATUS
U.S. 50 STATE
CURRENT LIMITS
1.2L O.C.P. ENGINE
1.8L G.M. WORLD CAR
0.33
.411
0.76
3.40
0.36
0.40
93
104
HC (g/mile)
CO (g/mile)
NOx (g/mile)
FUEL ECONOMY (g/mile)

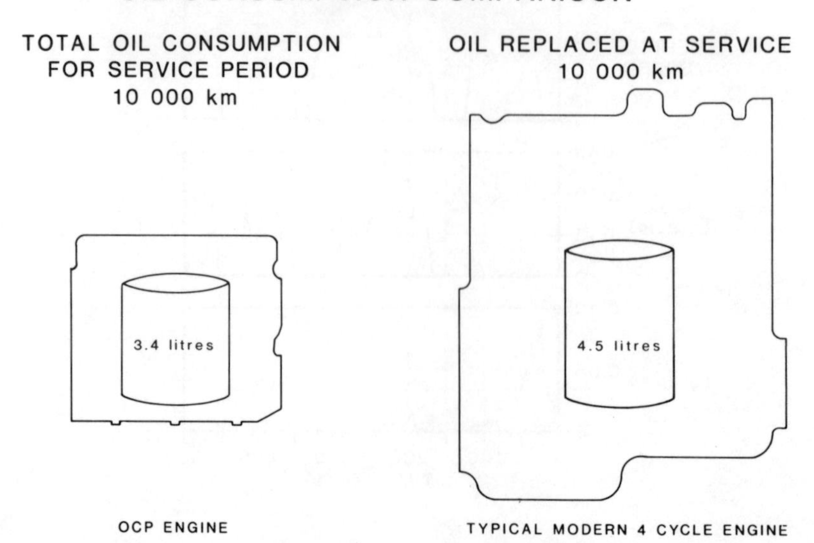
OCP ENGINE
OIL CONSUMPTION COMPARISON
TOTAL OIL CONSUMPTION
FOR SERVICE PERIOD
10 000 km
OIL REPLACED AT SERVICE
10 000 km
3.4 litres
4.5 litres
OCP ENGINE
TYPICAL MODERN 4 CYCLE ENGINE

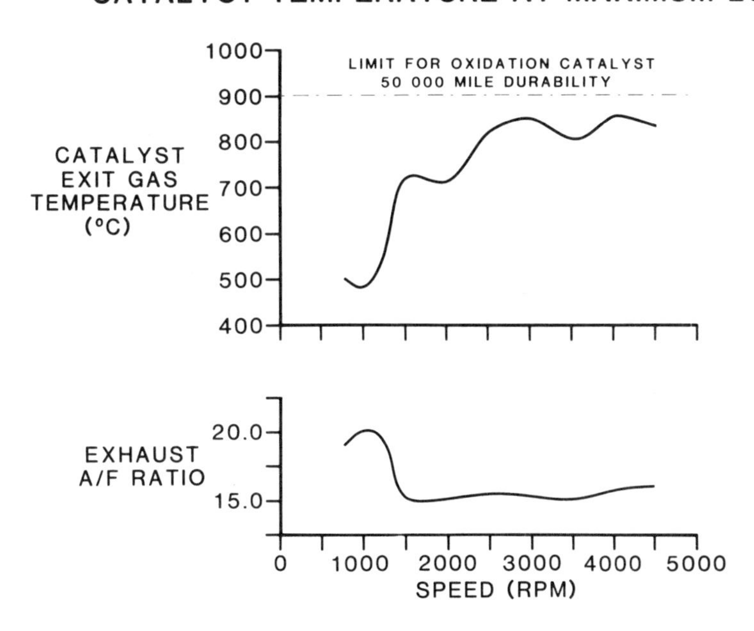
CATALYST TEMPERATURE AT MAXIMUM LOAD
LIMIT FOR OXIDATION CATALYST
50 000 MILE DURABILITY
CATALYST
EXIT GAS
TEMPERATURE
(°C)
1000
900
800
700
600
500
400
EXHAUST
A/F RATIO
20.0
15.0
0
1000
2000
3000
4000
5000
SPEED (RPM)

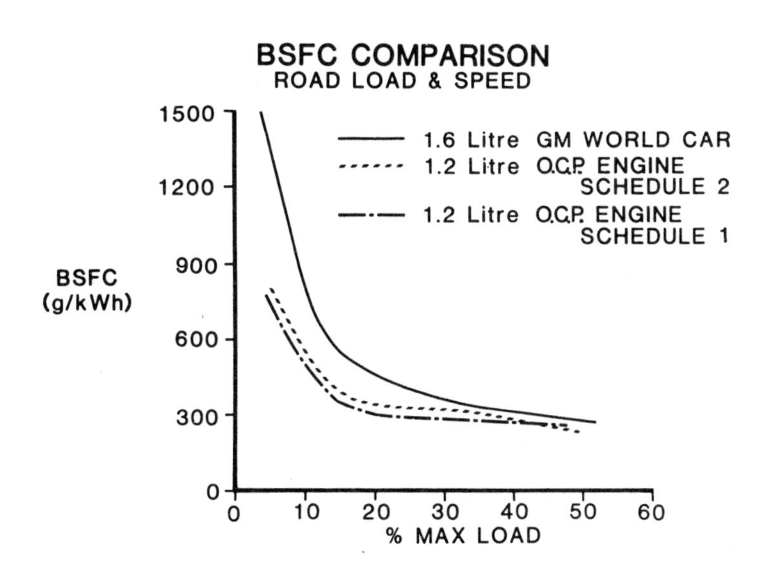
BSFC COMPARISON
ROAD LOAD & SPEED
1.6 Litre GM WORLD CAR
1.2 Litre O.C.P. ENGINE SCHEDULE 2
1.2 Litre O.C.P. ENGINE SCHEDULE 1
BSFC
(g/kWh)
1500
1200
900
600
300
0
10
20
30
40
50
60
% MAX LOAD

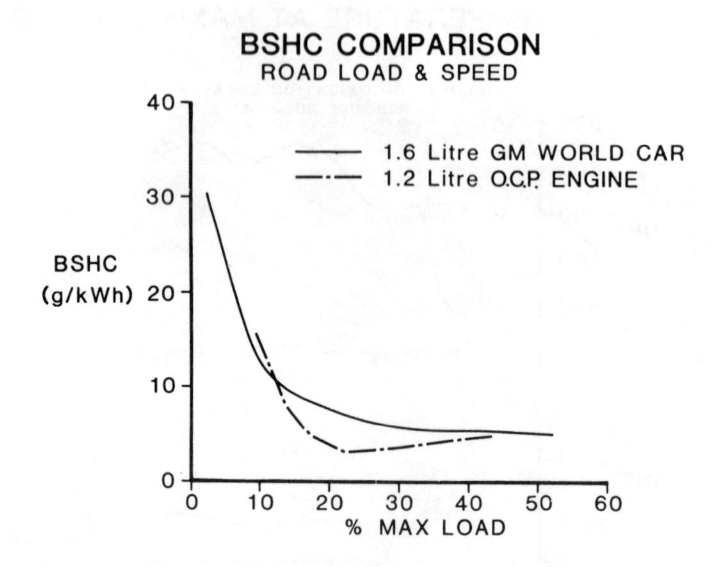
BSHC COMPARISON
ROAD LOAD & SPEED
1.6 Litre GM WORLD CAR
1.2 Litre O.C.P. ENGINE
BSHC
(g/kWh)
40
30
20
10
0
0
10
20
30
40
50
60
% MAX LOAD

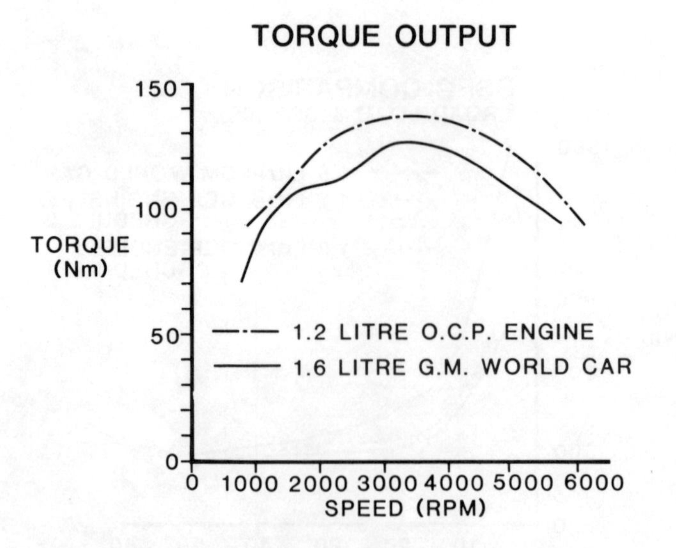
TORQUE OUTPUT
150
100
50
0
TORQUE
(Nm)
1.2 LITRE O.C.P. ENGINE
1.6 LITRE G.M. WORLD CAR
0
1000
2000
3000
4000
5000
6000
SPEED (RPM)

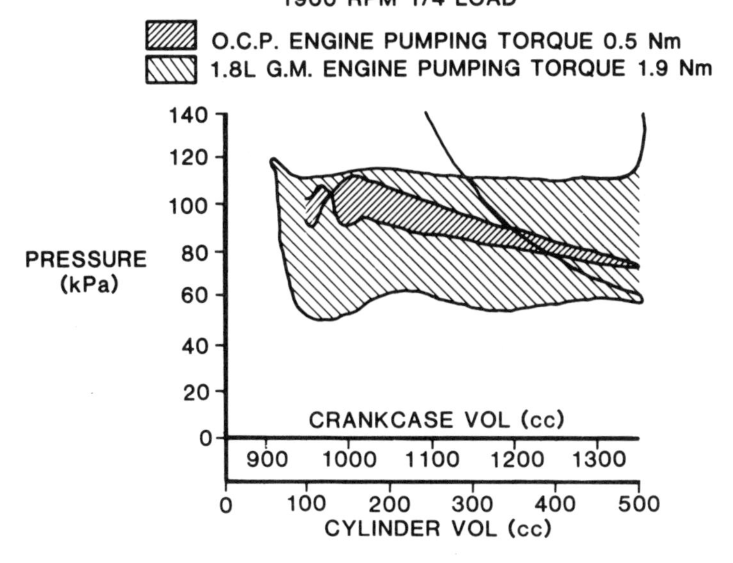
PUMPING WORK COMPARISON
1900 RPM 1/4 LOAD
O.C.P. ENGINE PUMPING TORQUE 0.5 Nm
1.8L G.M. ENGINE PUMPING TORQUE 1.9 Nm
PRESSURE
(kPa)
140
120
100
80
60
40
20
0
CRANKCASE VOL (cc)
900
1000
1100
1200
1300
0
100
200
300
400
500
CYLINDER VOL (cc)

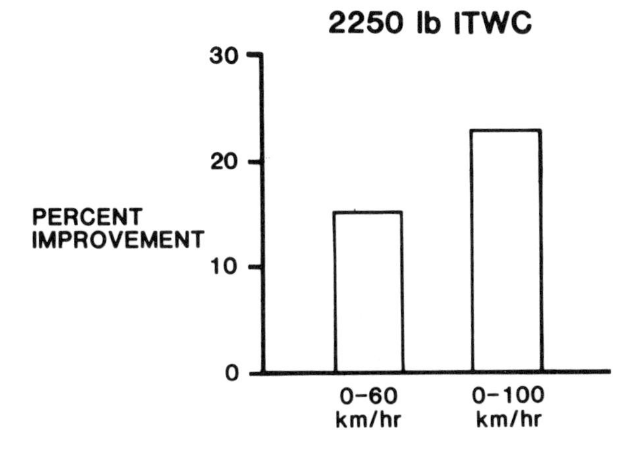
O.C.P. VEHICLE
ACCELERATION IMPROVEMENT
2250 lb ITWC
PERCENT
IMPROVEMENT
30
20
10
0
0-60
km/hr
0-100
km/hr

Index

NB: As this book largely concerns Ralph Sarich, the Orbital engine, the OCP engine and the Orbital Engine Company (OEC), separate listings are too numerous to include in the index.